FLYING SCOTSMAN

THE WORLD'S MOST FAMOUS LOCOMOTIVE

FLYING SCOTSMAN

BRIAN SHARPE

First published as Flying Scotsman: A Legend Reborn in 2016
by Mortons Media Group Ltd.

This edition published in 2019 by Gresley Books,
an imprint of Mortons Books Ltd.
Media Centre
Morton Way
Horncastle LN9 6JR
www.mortonsbooks.co.uk

ISBN 978-1-911658-02-3

The right of Brian Sharpe to be identified as the author of this work has been
asserted in accordance with the Copyright, Designs and Patents Act 1988.

Typeset by Kelvin Clements
Printed and bound by Gutenberg Press, Malta

10 9 8 7 6 5 4 3 2 1

From British Empire exhibit to
National Railway Museum treasure...

CONTENTS

Introduction

A LEGEND REBORN

✳✳✳

WHY DOES the British public hold one particular steam loco-motive in such high regard? Of the thousands of steam engines designed and built in Britain, the one that every-one knows and loves is *Flying Scotsman*. Several times since 1963, the engine has nearly been lost: either scrapped, exiled across the Atlantic or consigned to an inactive future on static display. But each time it gets another reprieve and returns to grace Britain's main lines once again, its return is even more spectacular than the time before and the British public turns out in their tens of thousands to welcome it back.

When the National Railway Museum finally saved it for a grateful nation in 2004, it looked as if *Flying Scotsman* would quickly be in steam, travelling the country and entertaining its audience. Yet it took £4.2m and ten years of heartache before a comeback on February 25, 2016, heading an express train once more on the line it was built for, from King's Cross to York.

It was a close-run thing, but all was well with the engine on the day. No other steam engine has the history of *Flying Scotsman,* and none has such a memorable name. This is the story of the engine and the train it once hauled, right up to the present day when *Flying Scotsman* is once again spending a busy few years doing what it does best.

No. 1472 was the third of a class of steam locomotives that was eventually to number 78 engines, and did not originally even carry a name. The Great Northern Railway A1 4-6-2 though, was the biggest express steam engine ever to have been seen in Britain at the time and was chosen to be displayed at a major exhibition at Wembley in 1924, for which it was given the name *Flying Scotsman*.

It hauled the London & North Eastern Railway's first King's Cross-to-Edinburgh nonstop express in 1928 and was the first steam engine in the world to officially break the 100mph barrier in 1934 although unofficially this speed had been achieved 30 years earlier. An identical engine to *Flying Scotsman* soon eclipsed its record with a 108mph burst of speed in 1935.

Flying Scotsman was perhaps becoming the best-known of its class, but none of its record feats actually stood for long. In 1935, Gresley's original Pacifics were superseded by the A4s: streamlined engines with more speed and power, and these raised the speed record first to 112mph, and later to 126mph.

From then on, *Flying Scotsman* was just one of many engines that played a vital part in hauling East Coast Main Line expresses between King's Cross, the north and Scotland, for another 30 years, but it had no more claim to fame than any of the others.

It still had its name though, and, when the final curtain came in early 1963 and the engine was withdrawn from service by British Railways and expected to be scrapped, it was purchased by a businessman who had every intention of keeping the engine running.

Flying Scotsman was certainly well known from the early years of its main line career, but it was 1963 when it really started to hit the headlines — after it had retired. This might not have happened had the engine not had such a memorable name.

Flying Scotsman has now become the one steam engine in the world of which many people know the name, and that most people would recognise. It was briefly the only main line steam engine running in the whole of Britain, and it has travelled across the Atlantic and across

America. It has circumnavigated the globe, steamed across Australia, broken the record for a nonstop run with steam (again), and been sold for easily the highest price ever paid for a steam engine.

But it has had its downsides, too. It has had several owners, some of whom have bought it on the strength of its earnings potential, based on the name *Flying Scotsman*. This value has perhaps been overestimated, and two of *Flying Scotsman*'s one-time owners have been bankrupted. Even the National Railway Museum has had its fingers burnt, but that is now history; *Flying Scotsman* is back!

It has been said that the engine's fame is such that it should have been preserved by the nation anyway. A large number of steam engines were preserved 'officially' and many are now on display in the National Railway Museum at York, but *Flying Scotsman* was simply not considered unique or historically important enough at the time to be included.

In 2004 though, its ongoing 40 years of fame (if not fortune) finally earned it a place in the National Railway Museum collection. After an unprecedented fundraising campaign and a National Heritage Memorial Fund grant, the museum was able to clear the enormous debts of the engine's owners and acquire *Flying Scotsman* for the nation, and for a British public that clearly holds the engine in high esteem.

If it had not acquired fame, largely as a result of its name, in the 1920s and 1930s, then maybe Alan Pegler would not have had the enthusiasm to purchase it in 1963. If it had been scrapped, what would then have become Britain's most famous steam engine?

The question is whether *Flying Scotsman* can run forever. The answer is probably yes, at a price. Like any steam engine, it is a mechanical object, built of steel. As parts wear out, they are replaced. Little of the original engine now exists and there has been much rebuilding and improvement carried out, before and after 1963.

And now it's back. No one expected it to take ten years of toil, but in 2016 the National Railway Museum finally had an engine to be proud of. The legend that is *Flying Scotsman* can run for ever, recognisable as the ultimate in British express steam design elegance. It may not be all the original steel, but the legend goes far beyond the engine's physical characteristics.

Chapter 1

THE GREAT NORTHERN RAILWAY

✳✳✳

Britain's first public railway was the Stockton & Darlington, which opened in 1825. The Liverpool & Manchester moved the story on in 1830, but the early independent railways all had one thing in common: they had quite unimaginative names, such as the Bristol & Exeter Railway. As they grew bigger though, they coined names like the North Eastern or the Midland, and as they grew even bigger, the prefix 'Great' became popular, with the Great Western being quite an apt description of that railway, while the Great Northern was really a bit of an exaggeration. It was neither great in the way that the GWR covered the whole of western England, nor was it really northern, as its main line to the north ran only as far as just beyond Doncaster.

The Great Northern Railway had as its aim the building of a railway from London to York, by the shortest possible route. It was relatively late in setting out to achieve this, and other companies' trains were already running between the two cities, but by a very circuitous route through Rugby and Derby.

The Act of Parliament authorising the construction in 1846 actually included two different routes, and the first section to be started was the line from Peterborough to Lincoln via Boston. In 1850 the line south was opened from Peterborough to a temporary London terminus at Maiden Lane, with the better-known King's Cross station, located in a then very disreputable part of London, opening in 1852.

Lincoln to Doncaster via Gainsborough was not completed until 1867 so for a short period, the GNR's expresses to the north ran via Boston to Lincoln, then over the Manchester Sheffield & Lincolnshire Railway to Doncaster via Retford. So much for the direct route to York! By the time King's Cross opened though, the GNR's 'towns line' from Peterborough to Doncaster, via Grantham, Newark and Retford, had been completed.

The GNR never reached York; it made it only as far as Askern Junction, just north of Doncaster, from where its trains ran over Lancashire & Yorkshire Railway tracks to Knottingley, then over part of the North Eastern Railway to York. It was not until as late as 1871 that the NER built a more direct line south from Selby to meet the GNR at Shaftholme Junction just a few yards from Askern Junction, completing what we now know of as the East Coast Main Line from King's Cross to Edinburgh. One of the principal Anglo-Scottish expresses was to become known, though unofficially, as the 'Flying Scotsman'.

Competition was fierce, especially for Anglo-Scottish passenger business in the 19th century. Trains ran from King's Cross, the GNR terminus, and Euston, the LNWR terminus, via the east coast and west coast routes respectively, beyond Edinburgh to Aberdeen. The east coast partners of Great Northern, North Eastern and North British railways, and their west coast opposite numbers, the London & North Western and Caledonian railways staged the 'Races to the North' in 1888 and 1895.

Despite the route being longer and hillier, the west coast alliance won the race both times. But it was the GNR that tended to be the senior partner in the east coast collaboration and, although it was on the losing side, its contribution should not be underestimated. The track on its main line was some of the best in the country. Its engines were among the fastest and most reliable, and its coach designs were adopted as standard for the 'East Coast Joint Stock' fleet used on Anglo-Scottish services. Despite the public relations 'spin' that its choice of company name implied, the Great Northern was one of Britain's railway companies that rarely gave any of its engines names. Perhaps its best-known 19th-century engine, the pioneer Stirling 8ft Single, is simply known as No. 1.

The numerous independent companies bearing the well-known names of the late 19th century, which built Britain's railway system, were to amalgamate into larger organisations in 1923, when the Grouping was to form the Big Four.

The GNR had main lines to Grimsby through the east Lincolnshire main line and to Leeds and Bradford in the West Riding of Yorkshire, plus branches serving Cambridge and Nottingham. The company also carried very heavy coal traffic south from Yorkshire, much of it later running over the line owned jointly with the Great Eastern Railway from Lincoln to March. Today, apart from the main line north from King's Cross, little of the 1,051 miles of the GNR route survives.

Most of its Lincolnshire lines closed in 1970, apart from the Skegness branch. The Peterborough-to-Doncaster-via-Lincoln line sees an hourly one-coach train, but the branches from the main line still serve Cambridge and Nottingham. Most ECML expresses are now operated by London North Eastern Railway but services are also run by Grand Central and by Hull Trains.

It was the Great Northern Railway's public relations exaggeration in terms of its name, where the myth appeared to be more important than the reality, that really set the scene for the building of the world's most famous steam engine.

Chapter 2

HERBERT NIGEL GRESLEY

✳✳✳

ERBERT NIGEL Gresley was born on June 19, 1876 at 14 Dublin Street, Edinburgh, the youngest of five children born to Reverend Nigel Gresley and his wife, Joanna. After leaving Marlborough College, the young Gresley was taken on as an apprentice by the London & North Western Railway at Crewe, on his 17th birthday, working under the much-feared Francis Webb. Gresley was described as "more ponderous than talented" at Crewe, but he was certainly ambitious.

In 1898, he moved on to the Lancashire & Yorkshire Railway, where he first worked in the drawing office. There were unconfirmed reports that, in that year, a Lancashire & Yorkshire Railway steam engine, 4-4-2 No. 1352, allegedly exceeded 100mph on a run between Liverpool and Southport. These types of engines were known as 'Highflyers' and the young Gresley certainly seems to have been a high-flyer himself. In 1900 he was appointed running shed foreman at Blackpool, and soon afterwards promoted to assistant manager at Newton Heath carriage works in Manchester.

Gresley married Ethel Frances Fullagar in 1901. In 1902, he was promoted again, to works manager at Newton Heath, doubling his £250 annual salary. In 1904 he was promoted again to carriage and wagon superintendent of the LYR in the year that the somewhat more plausible 102.3mph run by the Great Western Railway's 4-4-0 No. 3440 *City of Truro* took place on Wellington Bank in Somerset. Although never totally authenticated, it is generally felt that *City of Truro* at least came very close to achieving 100mph, and was the fastest man-made machine in the world at the time.

The Lancashire & Yorkshire was certainly a well-respected railway, but its trains did not quite have the prestige of the Anglo-Scottish expresses. After an interview with its locomotive engineer, H.A. Ivatt, and on the recommendation of John Aspinall of the LYR, Gresley joined the Great Northern Railway as assistant carriage and wagon superintendent in January 1905.

He quickly made his mark on the GNR. His design of elliptical-roofed, steel-framed and teak-bodied coaches was introduced within a year and became standard for the GNR and for East Coast Joint Stock for the Scottish expresses. The coaches set new standards, many features becoming adopted for coach design for many years to come, and the last ones were still in use on British Railways' express trains into the mid-1970s. Gresley's design of coach bogie was to last even longer, being used on Southern Region electric stock well into the 21st century.

Gresley's coaches were quickly to replace the four- and six-wheeled clerestory-roofed coaches still in widespread use on the GNR. He also made innovations, such as electric lighting and articulated coaches, first for suburban services, but much later even for express trains.

In 1911, at the age of 35, Gresley succeeded Ivatt as the GNR's locomotive engineer, and the railway industry waited to see whether he would be as forward-thinking and innovative with his locomotive designs as he had been with his carriages. Ivatt was a well-respected engineer, and his engines were good, but at the time they were not the best in Britain. Ivatt had recommended Gresley as his successor, but the GNR board must have had reservations. Gresley was a confident, self-assured young man, who was ambitious and knew what he wanted.

His early locomotive designs for the Great Northern Railway were traditional but, when he started designing new express engines, they were revolutionary, especially the A1 class Pacifics of 1922. They were big, they were successful and they were noticed, not just by his contemporaries but by the press and the public. None, of course, caught the attention to quite the same extent as did the third of the A1 Pacifics, *Flying Scotsman*, but that was after the GNR had ceased to exist.

The responsibilities associated with the top job in the locomotive department of a railway company tended to change through the years; for example, on the GNR, Benjamin Cubitt was responsible for obtaining and maintaining the GNR's first locomotives and he was succeeded by Edward Bury but neither actually designed the company's engines or supervised their construction, although the railway did build locomotive workshops to carry out overhauls. It was only under Archibald Sturrock that the GNR built its own engines in its workshops, and Sturrock was credited with their design; the job title was Locomotive Superintendent.

The job evolved to include rolling stock and the title also evolved into that of Chief Mechanical Engineer. The CMEs did not physically design every part of the locomotives they are credited with designing, of course, although some, such as Nigel Gresley, had greater involvement than others in the finer details. They relayed the requirements of the management and operating departments to the design staff, and outlined the specifications and critical dimensions of the locomotives. The external similarity between the designs of one CME and those of his predecessor arises because the same team of designers would continue under the new leader.

Locomotive engineers were a breed apart; not only did they tend to move from one railway to another, but many were related by birth and several also became in-laws as marriages took place.

One example is Patrick Stirling. He was the cousin of his predecessor on the GNR, Archibald Sturrock. He had moved from being CME of the Glasgow & South Western Railway; his brother James Stirling was also once CME of the GSWR and at one time CME of the South Eastern Railway, and Patrick's son, Matthew, was a long-serving CME on the Hull & Barnsley Railway.

Dugald Drummond joined the London & South Western Railway from being CME of the Caledonian Railway. His brother, Peter, was CME of the Glasgow & South Western Railway.

S.D. Holden of the Great Eastern Railway was succeeded by his son as CME. H.A. lvatt's eldest daughter married O.V.S. Bulleid, Gresley's assistant on the LNER, who later became CME of the Southern Railway. Meanwhile lvatt's son became CME of the London Midland & Scottish Railway in 1945. Edward Thompson, who succeeded Gresley as CME on the LNER, was the son-in-law of Sir Vincent Raven of the North Eastern Railway.

The Chief Mechanical Engineers have tended to be the best-known names from the steam era among railway enthusiasts, far better known, for example, than the railways' chairmen or architects, but few CMEs names are known to the public at large. (See Appendix 1.)

GRESLEY'S GNR LOCOMOTIVE DESIGNS

There is no doubt that the GNR had some of Britain's finest 19th-century locomotive engineers in charge of designing its engines – Archibald Sturrock, Patrick Stirling and H.A. Ivatt were all in the premier league of their day.

In 1905, though, Herbert Nigel Gresley joined the company, initially being appointed as carriage and wagon superintendent. The GNR at the time was noted for still running some of the most ancient coaches of any major railway, mostly of six-wheeled design, but Gresley soon changed that.

Having been promoted to Chief Mechanical Engineer in 1911, Gresley finally became responsible for designing the company's steam locomotives and his first steam design for the GNR was the H1 (later LNER K1) 2-6-0 in 1912. They were conventional engines, in traditional GNR style, and designed to accelerate long-distance goods traffic. As early as 1915, though, Gresley was thinking about a four-cylindered Pacific for express trains; unfortunately, wartime was not a good time to build such a thing.

In 1918, Gresley designed the O1 2-8-0 for heavy freight traffic, which not only had three cylinders but also Gresley's new design of conjugated valve gear. This was to set the scene for Gresley's locomotives in the future, but he had yet to build an express engine. Partly this was because Ivatt's Atlantics were so good and were held in such high regard by Gresley that he felt he had little need to improve on them immediately, but it was also because the GNR was more in need of modern freight power.

When Gresley's first express engine did appear, in 1920, the H4 was a 2-6-0 — not a wheel arrangement normally associated with express power, particularly on level tracks. It was built for sustained power output, not high speed, with three cylinders and six relatively small driving wheels, which gave room for a huge boiler. The class was to be reclassified K3 by the LNER and these engines were still pulling summer Saturday expresses to the east coast in the early 1960s, by which time they were considered rather old and basic mixed-traffic engines. But when they were built, they were huge express engines that shocked everyone in the railway industry with their sheer size. Gresley was going for big engines, and was not universally popular with the GNR management because of what these monsters might do to track and bridges on the line. Undeterred, Gresley went for the biggest engine it was felt possible to build.

Stylistically, the similarity with the H4 (K3) was evident, but it was a 4-6-2, a Pacific, only the second to be built in Britain. The GWR had built the first, called *The Great Bear*, in 1908, and it had proved to be too big to be practical on that railway. Gresley had to compromise with his Pacific design; everywhere weight could be saved, it was, even to the extent of using a lower boiler pressure than would have been ideal.

The new engine, when it emerged from Doncaster works in 1922, carrying the number 1470, was considered so important that the GNR even chose to give it a name: *Great Northern*. Naming locomotives was contrary to the GNR's normal policy, but the exception was made to keep the company name alive, as the GNR was to disappear into the London & North Eastern Railway very shortly. But what had been an imaginative choice of name for the company was really rather less than inspiring for the locomotive.

A second engine followed, No. 1471, just before the Grouping of the railway companies on January 1, 1923.

CONJUGATED VALVE GEAR

The principle of conjugated valve gear, while not unique or exclusive to Gresley, was only ever used by him in any quantity of steam locomotives produced in Britain. Virtually all of his designs from the O1 2-8-0 onwards used this system, apart from smaller freight and shunting designs.

The valve gear on a steam engine is the complex arrangements of rods that can be seen connecting the cylinders to the driving wheels. It can admit steam to the cylinders at the right time for the engine to move forward or backward, slow or fast. In early engines such as *Locomotion*, it was outside (in fact, above) the engine. Later 19th-century designs tended to conceal the valve gear, if not the cylinders, inside, between the frames, for elegance. By the 1920s, though, outside cylinders and valve gear were becoming the norm.

Most engines were, of course, two cylinder anyway, but three- and even four-cylinder designs started to be adopted, which gave more even power when starting and at speed. In theory, these needed three or four sets of valve gear, with consequent additional weight and maintenance costs. Churchward on the Great Western opted for four cylinders, but used two sets of inside valve gear, controlling two cylinders each.

Gresley chose his three-cylinder design, but with only two sets of valve gear, outside the frames, but connected through the frames, and jointly controlling the inside cylinder.

On a conventional two-cylinder engine, the cranks on the driving wheels are set at 90° to each other, with each cylinder giving a forward and backward thrust; there are four thrusts at even intervals per revolution of the wheel.

Normally a four-cylinder engine has eight thrusts per revolution, but they are in pairs, so there is no major engineering problem in a set of valve gear controlling two cylinders that act in unison.

A three-cylinder engine, though, has its three cranks set at 120° to each other, so there are six thrusts per revolution. For two sets of valve gear to jointly control a third cylinder so that it acts identically to the one either side, but exactly midway between them, is quite an engineering achievement.

The arrangement had major advantages for smoothness, weight-saving and ease of maintenance, but was still more complex than a conventional two- or four-cylinder arrangement and, with many moving parts between the frames, adequate lubrication was critical. If a Gresley engine ever broke down, it was often caused by problems with the conjugated valve gear between the frames.

Chapter 3

PACIFICS

✳ ✳ ✳

*F*lying Scotsman, when built, was the third of 79 LNER Gresley
A1/A3 Pacifics. Designed by Gresley, it was built by the London
& North Eastern Railway, which classified it as an A1, but why
is it called a 'Pacific'?

As steam locomotives grew in size, they were inevitably designed with
more wheels, but not all were 'driving' wheels, i.e. directly powered by
the cylinders. George Stephenson's *Rocket* was the first steam engine
built for speed not power; i.e. the first passenger engine. In *Rocket*'s case,
only the leading wheels are powered, so the wheel arrangement of the
engine is described as an 0-2-2. The first digit is the leading (carrying)
wheels, the middle one the driving (coupled) wheels, and the third the
rear carrying wheels.

The more driving wheels an engine has, the more power it is able to
transmit to the track, but there are good reasons for not all of a locomotive's wheels being powered. The rigid wheelbase can be unkind to
the track, and there are more expensive bearings to maintain.

Eventually locomotives of a certain wheel arrangement started to
acquire nicknames. These clearly originated in North America, and tend
to refer to the areas where a particular wheel arrangement found favour.

The 4-4-2 'Atlantic' was built for speed on level track and was popular in the relatively flat states of the Eastern Seaboard. The 4-6-0 had more power but less speed and was useful in the Midwest, but its nickname, the '10-wheeler', never caught on in Britain. The 4-6-2 or 'Pacific' combined speed and power, useful on the West Coast, where hundreds of miles of level track could suddenly end with a ferocious gradient into the Sierra Nevada mountains.

This was nothing to the terrain encountered by lines such as the Great Northern Railroad, and nothing less than a massive 4-8-4 was necessary to shift tonnages across the Rockies. These acquired the name 'Northern' and were of the type that eventually found universal favour across the US and Canada.

British express steam power consisted mainly of 4-4-0s, 4-4-2s and 4-6-0s, until the coming of the 4-6-2s in the early 1920s. Even the adoption of the transatlantic Atlantic and Pacific names as descriptions of the engine types probably owed a lot to East Coast Main Line (especially Great Northern Railway) marketing spin. Few other railways even had Atlantics, let alone Pacifics, and 'Pacific' had much more public appeal and glamour than '4-6-2' could ever have.

Many of the larger wheel arrangements were rarely used in Britain and, in fact, several, such as 4-6-4, 2-8-2 and even 2-6-2, were seen almost exclusively in Nigel Gresley's later designs.

The Whyte system of wheel arrangements was not universal. France referred only to axles, so a 4-6-2 was a 231. Turkey used driving axles and total axles, so a 2-8-0 was a 450. Initially, diesels used the Whyte system, but as powered bogies became the norm, Bo-Bo and Co-Co came in to use; B being a two-axle bogie, and C three-axle, the 'o' denoting both being powered. An A1A-A1A has two three-axle bogies but with the centre one of each unpowered. Overleaf are some examples:

0-2-2	e.g. Stephenson's *Rocket*	Oo
0-4-0	e.g. *Locomotion No. 1*	OO
0-4-2	e.g. LMR *Lion* or LBSCR *Gladstone*	OOo
4-2-2	e.g. GNR Stirling 'Single' No1	ooOo
2-4-0	e.g. LNWR No. 790 *Hardwicke*	oOO
4-4-0	the classic Victorian passenger engine	ooOO
4-4-2	'Atlantic' as used by the GNR	ooOOo
0-6-0	the classic standard goods engine	OOO
2-6-0	'Mogul'	oOOO
4-6-0	'10-wheeler'	ooOOO
2-6-2	'Prairie'	oOOOo
4-6-2	'Pacific'	ooOOOo
4-6-4	'Hudson'	ooOOOoo
0-8-0		OOOO
0-8-2		OOOOo
2-8-0	'Consolidation'	oOOOO
2-8-2	'Mikado'	oOOOOo
2-8-4	'Berkshire'	oOOOOoo
4-8-2	'Mountain'	ooOOOOo
4-8-4	'Northern'	ooOOOOoo
0-10-0	'Decapod'	OOOOO
2-10-0	e.g BR 9F such as *Evening Star*	oOOOOO
2-8-8-2	Garratt, just one example of articulated locomotive designs with separate sets of coupled driving wheels.	oOOOO OOOOo

Chapter 4

THE GROUPING

✳✳✳

ON JANUARY 1, 1923, Britain's independent railway companies merged to form what became known as the Big Four: the Great Western Railway, the Southern Railway, the London Midland & Scottish Railway and the London & North Eastern Railway.

The GWR, in fact, was relatively unchanged, apart from absorbing the various smaller independent companies in south Wales and the Cambrian Railways. The Southern Railway was a fairly simple amalgamation of the South Eastern & Chatham Railway, the London Brighton & South Coast Railway and the London & South Western Railway.

The LMSR principally combined the Midland Railway, the London & North Western Railway and, in Scotland, the Caledonian Railway, the Highland Railway and the Glasgow & South Western Railway. The Lancashire & Yorkshire Railway and the North London Railway had already merged into the LNWR, and the LMSR also took in the North Staffordshire, Furness and Maryport & Carlisle railways.

The LNER was an amalgamation of the Great Northern, Great Central, Great Eastern and North Eastern railways in England, and the North British and Great North of Scotland railways north of the border. The Hull & Barnsley Railway had already merged with the

NER, but the Midland & Great Northern Joint Railway was to remain independent until 1936.

The effect of the Grouping on the *Flying Scotsman* story was twofold. The train, the 10am from King's Cross and corresponding Up working from Edinburgh, was now run by one company throughout, as was the competing train on the West Coast Main Line. The engine, No. 1472, under construction at Doncaster, and which would shortly become *Flying Scotsman*, although built to run on the Great Northern Railway between King's Cross and York, was now likely to run much farther afield, perhaps to Edinburgh or even beyond.

GNR/LNER LOCOMOTIVE CLASSES

The engine types of the Great Northern Railway were reclassified by the LNER although, where possible, the same ones were retained, and the other constituent companies' classes altered to fit.

The GNR had rightly called its new Pacifics A1s and this was continued. On the LNER, the letter denoted the wheel arrangement. The number within the wheel arrangement was generally in order of the constituent companies. So, the GNR Pacifics were A1s and the NER Pacifics A2s.

As classes were rendered extinct by withdrawals, their numbers were taken by new LNER designs, so we saw Thompson and Peppercorn A1 Pacifics, Peppercorn A2s, Thompson B1s and Peppercorn K1s in later years. As far as possible, the GNR classifications were used, but the opportunity was taken to make the system more logical. (See Appendix 2.)

Chapter 5

'FLYING SCOTSMAN' AND *FLYING SCOTSMAN*

✳✳✳

'THE FLYING SCOTSMAN': THE TRAIN

THE NAME 'Flying Scotsman' had unofficially been given to the premier Anglo-Scottish express train from King's Cross, the 10am to Edinburgh, a service that dated right back to Great Northern Railway days in 1864.

At that time, the train was the fastest on the East Coast route — allowed, for example, just 1 hour 35 minutes to cover 76.4 miles from King's Cross to Peterborough, a time that remained relatively unaltered throughout steam days. Arrival at Edinburgh was at 8.30pm, a time that also remained unaltered for many years, but gradually moved forward as locomotives became more powerful.

The GNR was responsible for operating the train as far as York, and stops were made at Peterborough, Grantham and Retford. For much of its earlier years, the 10am passed nonstop through Doncaster. In fact, for several years, a slip coach was included, which was detached on the move to serve Doncaster. For a few years another slip coach was detached at Essendine for Stamford passengers.

By 1910, the train was running from King's Cross as far as York with only one stop, at Grantham, and was reaching Edinburgh in 8hr 45min and, during the summer, a 9.50am relief ran, nonstop to Doncaster, then nonstop to Newcastle with a North Eastern Railway engine. The train was colloquially known as the 'Special Scotch Express', then 'The Flying Scotchman', said to have originated partly from the famous ghost ship The Flying Dutchman, but also as express stagecoaches were often referred to as 'Flying'.

However, officially, the GNR always referred to the train simply as the 10 o'clock. The LNER continued running the train and the new Gresley A1 Pacifies were an ideal choice of motive power from 1923. In 1927, the relief train started running nonstop from King's Cross to Newcastle, with A1 Pacific No. 4475 Flying Fox being the engine chosen to inaugurate what was to be one of the longest regular nonstop workings by a steam engine.

By then, the 'Flying Scotchman' had become known as the 'Flying Scotsman' and, from 1927, it became official, with its title appearing in timetables, on carriage roofboards and on a headboard carried on the engine.

The rival LMS was outdoing the LNER by running nonstop between Euston and Carlisle, using a Royal Scot 4-6-0, and this was slightly further than King's Cross to Newcastle. The two companies had a gentlemen's agreement not to compete for the fastest speeds on the Anglo-Scottish runs, partly for safety reasons: competition was in the form of style, publicity, passenger comfort and amenities such as bars, restaurants and even hairdressing salons.

The LNER on May 1, 1928 started to run its 10am 'Flying Scotsman' service nonstop from King's Cross to Edinburgh and, appropriately enough, the engine that hauled the first northbound train was No. 4472 Flying Scotsman itself. The launch of this service was a huge media event, and the 'Flying Scotsman' was seen by the sort of crowds that witness the engine's passage today. Times have changed in that the first run of a regular daily service could command such public and media interest.

The LMS actually beat the LNER a few days earlier by running nonstop from Euston to both Glasgow and Edinburgh. However, these were one-offs; the LMS could not expect its crews to run that kind of distance daily. The LNER, though, had fitted some of its Pacifics, including *Flying Scotsman*, with corridor tenders so the crews could be changed en route at the halfway point just north of York.

The 'Flying Scotsman' continued to run, almost without interruption, as other named trains on the route came and went. It was accelerated in 1932 for the first time in 32 years but, by 1938, the 'Coronation' was faster. From 1949, the 'Capitals Limited' became the Edinburgh nonstop service, renamed the 'Elizabethan' in 1953. Only with the introduction of the Deltic diesels in 1961 was the 'Flying Scotsman' accelerated to equal the pre-war steam times of the 'Coronation'. The 'Elizabethan' was never dieselised and was worked by Gresley A4 Pacifies to the end.

Some of the pre-war 'Silver Jubilee' and 'Coronation' streamliners, the post-war 'Capitals Limited' and 'Elizabethan', and the diesel-era 'Silver Jubilee' were faster trains; some ran nonstop to Edinburgh while the 'Scotsman' started to pause at Newcastle once more, but they were all shortlived by comparison.

The 10am 'Flying Scotsman' outlasted them all until 1982, when its departure time was changed to 10.35am.

FLYING SCOTSMAN: THE ENGINE

By the time Nigel Gresley's third A1 Pacific No. 1472 actually took to the rails, the Great Northern Railway had become part of the London & North Eastern Railway when the independent companies were grouped into the Big Four.

Gresley could actually have lost his job through the Grouping, as although Robinson of the Great Central was the preferred choice as the CME of the new company in view of his experience, he felt ready to retire and instead recommended Gresley for the job. The whole course of railway history in Britain could have been very different, and it was Gresley's appointment as the LNER's CME that led to production of his Pacifics being continued and expanded.

No. 1472 was painted in the LNER standard livery of apple green, a simpler version of GNR livery, with black and white lining. In a departure from previous practice, the number was painted on the huge eight-wheeled tender in large gold shaded lettering, with L&NER in smaller letters above it. Having inherited engines from several different companies, the LNER found itself with many engines with identical numbers, and part of the solution was to add 3,000 to all GNR numbers, so the third A1 quickly became No. 4472.

The LNER had named the second Pacific, No. 1471 *Sir Frederick Banbury* but No. 1472 and subsequent engines remained unnamed. When it was decided to send the third Gresley Pacific for display at the British Empire Exhibition of 1924 in Wembley, it was given the name *Flying Scotsman*, the name by which the GNR's premier express train had been unofficially known.

The LNER's decision to exhibit this engine at the British Empire Exhibition at Wembley in 1924 led to other major changes in No. 4472's external appearance. The LNER coat of arms was added to the cabsides, the wheel tyres and splasher beadings were polished metal and, most importantly, the *Flying Scotsman* nameplates were fitted over the centre driving wheel splasher.

Flying Scotsman's appearance was to change quite regularly, sometimes to a limited extent but sometimes quite drastically over the next 93 years. When first built, the A1s were too big for the turntable at King's Cross, and although the Grouping opened up a much wider sphere of operation, the Pacifics needed a bit of trimming down, so slightly lower chimneys and domes were fitted and the bottom corners of the bufferbeams were cut away. This now enabled the A1s to reach Edinburgh and Aberdeen without striking any bridges or platforms.

But were Gresley's A1 Pacifics any good?

Well, at the same time as Gresley's engines rolled out of Doncaster, Sir Vincent Raven of the North Eastern Railway was producing engines of a new Pacific design at Darlington. These were known as A2s on the LNER and the two types were tested against each other, just to prove to Gresley that his design was best and there was nothing he could learn from Darlington. Raven was a respected engineer and his engines were good, but Gresley's were considered better so the Raven design only ever totalled five engines while production of the Gresley ones continued, to eventually total seventy-nine.

When *Flying Scotsman* appeared at the British Empire Exhibition at Wembley in 1924, it stood next to the GWR's 4-6-0 No. 4073 *Caerphilly Castle*, which the GWR's publicity described as Britain's most powerful express engine. To prove it, GWR general manager Sir Felix Pole proposed to the LNER that an exchange trial of the two types should take place. In April 1925, *Pendennis Castle* ran on the Great Northern main line and LNER A1 No. 4474 *Victor Wild* represented the LNER on the GW.

It is generally accepted that the GWR locomotive won the battle in terms of speed, power and especially economy and this led to improvements being made to the LNER Pacific design. The GWR rather rubbed salt in the wound by sending *Pendennis Castle* fresh from its triumphs, to the British Empire Exhibition which had reopened for the summer of 1925, where *Pendennis* stood alongside *Scotsman*, carrying a headboard proclaiming the Castle to be Britain's most powerful locomotive. However, despite *Pendennis Castle*'s superiority in the contest, the name was just never going to stick in people's minds quite like *Flying Scotsman*. Interestingly for the 1925 exhibition, No. 4472 was coupled to a standard LNER-size-wheeled tender.

Gresley had little direct involvement in the contest but took the lessons on board, in particular the use of long-lap piston valves. This, together with increasing the boiler pressure from 180 to 220psi, increasing the amount of superheating, and slightly reducing cylinder diameter, transformed the A1 Pacifics; those built to the new specification were classified A3.

Flying Scotsman had the name that made it a natural choice for any publicity-oriented stunt the LNER wanted to stage, so it quite naturally took part in the 1929 publicity film, *Flying Scotsman*, but did not haul the inaugural nonstop train to Newcastle in 1927. It did, though, haul the rather more important first nonstop to Edinburgh on May 1 the following year.

It was also one of the engines chosen to take part in some high-speed test running in 1934, in the course of which it not only broke the 100mph barrier but did it while hauling a train carrying scientific equipment to record its speed. It may not really have been the first steam locomotive to reach 100mph, but it was definitely the first one officially and scientifically recorded as having done so. The name *Flying Scotsman* on the engine just always seemed to attract rather more media coverage for these exploits than might have been the case with any other engine.

It was No. 2750 *Papyrus* that hit 108mph soon afterwards, but how many people have ever heard of it?

If Gresley had not been appointed Chief Mechanical Engineer of the LNER, Robinson, or possibly someone else, would have been appointed, and the new Pacifics could have had a much less certain future. Even if the new man had not favoured Gresley's more radical design features such as conjugated valve gear, no one could have rebuilt the A1s or come up with an alternative Pacific design in time for the 1924 Wembley Exhibition.

An A1 would almost certainly still have been selected and the name *Flying Scotsman* is likely still to have been carried. *Flying Scotsman*'s initial claim to fame would still have been there and the combination of the most elegant and powerful-looking Pacific and the most inspired choice of name bestowed on a steam engine would still have assured it a place in railway history.

However, it could have been shortlived. A new CME might not have persisted with the design and acted on the results of the comparative trials with the GWR Castle, as did Gresley. The A1 was not actually as good as it looked initially, and a new CME might have simply thought he could do better and come up with his own new design, which would have been selected for the record nonstop runs and speed trials.

Other steam designs elsewhere proved that they could attain speeds of well over 100mph, and it was the limitations of the routes they ran on, rather than any design weakness, that prevented them from achieving *Mallard*'s 126mph record.

Another designer's engine might well have attained the steam speed record for the LNER, and *Flying Scotsman*'s fame would then have been brief. It might even have remained one of a handful of engines, such as Raven's LNER A2 Pacifics, to be scrapped in the 1930s. Perhaps the name would have been reused and *Flying Scotsman* might still have gone on to become the most famous steam engine in the world, albeit a completely different engine.

All this is not just conjecture, but unlikely. Sir Nigel Gresley was one of the greatest engineers ever to have lived and his achievements were already so prolific by 1922 that his appointment as CME by the LNER was virtually assured.

Chapter 6

WHAT'S IN
A NAME?

✳✳✳

IMPORTANT EXPRESS trains have been named since the early days
of long-distance train travel, following the traditions of the days of
stagecoaches but, in most cases, initially the name was not officially
given to the train by the railway that ran it.

Britain is actually behind other countries in naming its trains, but this
is partly because of the sheer volume of trains that run in this country.
In the United States, many lines see only one passenger train per day,
so naturally it is named.

Today, many of the famous named trains, as opposed to locomotives,
are well-known to most people, although in most cases, in the era of
standardisation and modern traction, the names are rarely now used.

It is sometimes difficult to establish exactly when a train received its
name; certainly, it is very rare to see photographs of pre-Grouping trains
with headboards carried on the locomotive. In many cases, the train
name seems to have been used mainly in advertising and for publicity,
especially for summer holiday trains. A classic example is the 'Irish
Mail', a title used right from the first train from Euston to Holyhead

connecting with the boat to Ireland in 1848, but not distinguished officially even by carriage roofboards until 1927.

The Great Eastern Railway half-heartedly named one train in 1897 the 'Cromer Express' and, when it changed it in 1907 to the 'Norfolk Coast Express', did provide it with carriage roofboards but no locomotive headboard.

The names used in the early part of the 20th century reflect the era when language was different to today. One of the first named trains to run in Britain appears to have been the 'Brighton Sunday Pullman Limited', from around 1899. In 1908 it became the slightly more modern-sounding 'Southern Belle' and the 'Brighton Belle' from soon after electrification in 1932.

At the other end of Britain, Scottish expresses to Aberdeen had been named from 1906, including the Caledonian Railway's 'Granite City' and 'Grampian'. North British Railway locomotives normally carried a destination board on the front of the engine on most passenger services, so this company tended to be at the forefront of the tradition of also identifying its named trains with a locomotive headboard, including the 'Fife Coast Express' from 1910 and the 'Lothian Coast Express' from 1912.

The GER and NBR named trains were early casualties of the First World War, but the NBR trains were quickly revived afterwards so the LNER did inherit some named trains, and no doubt the NBR policy was to influence the LNER's decision to name its premier long-distance expresses soon after Grouping in 1923.

The Great Western Railway was certainly a world leader in the early 1900s and its 10.10am Paddington-to-Penzance express ran nonstop to Plymouth from 1904, easily the longest nonstop run at that time. By 1906 it was known as the 'Cornish Riviera Limited', a train that has continued to run almost uninterrupted, apart from a slight name change to 'Cornish Riviera Express'. The word 'Limited' in a train name is typical railway parlance and refers to the fact that the train was booked to make only a limited number of stops. These names were hardly compatible with modern marketing-speak.

Another early named train was the 'Sunny South Special', later 'Express', initially a joint London & North Western and London Brighton & South Coast Railway operation, but later extended to a series of Midlands and

North, to South coast summer holiday trains. It is doubtful whether any was ever distinguished with a headboard or carriage roofboards, though.

From 1923, the GWR's 'Cheltenham Spa Express' officially became Britain's fastest train and was known as the 'Cheltenham Flyer', which officially it never was. Again, the public knew the name it wanted, but the railway was too staid to listen to public opinion.

Certainly, a very early British train to have acquired a name was the 10am King's Cross-to-Edinburgh, referred to as the 'Special Scotch Express' in timetables from 1864 but, from the 1880s, referred to by the public as the 'Flying Scotchman' and later still 'Flying Scotsman'. It was the public that named it though, not the railway operating authorities until 1924. It was in 1926/7 that the LNER recognised there was much to be gained by listening to public opinion, and it not only officially recognised the name 'Flying Scotsman', already by then carried on A1 Pacific No. 4472, but, at the same time, the 'Aberdonian', 'Night Scotsman' and 'Scarborough Flyer' names were given official recognition, followed by the 'Queen of Scots' Pullman in 1928. These trains were distinguished by the carrying of headboards on the locomotives and roofboards on the carriages.

The railway publicity machines started working to good effect in 1927, and the name 'Flying Scotsman' was simply the best of a great selection of train names that were either updated, officially adopted or dreamed up at around this time.

The LNER was not unique and not necessarily the trendsetter. At the same time, the Southern Railway's 'Atlantic Coast Express' was named, as was the GWR's 'Cambrian Coast Express' and the Manchester-to-Bournemouth 'Pines Express'.

The LMSR naturally countered the LNER publicity machine with its 'Royal Scot' and 'Royal Highlander' Anglo-Scottish expresses, yet the carrying of headboards on engines never became standard practice on the LMSR.

On the Southern Railway, the international express from Victoria to Dover became the 'Golden Arrow' in 1929, with its French counterpart carrying the 'Flèche d'Or' insignia. The 'Bournemouth Belle' Pullman followed in 1931.

Surprisingly, the GWR's 'Bristolian' did not come into being until 1935, at around the time the real publicity and speed competition started, with the LNER's 'Silver Jubilee' and 'Coronation' being countered by the LMS 'Coronation Scot'.

During the Second World War, though, only four trains retained their names: the East Coast Main Line's 'Flying Scotsman', 'Night Scotsman' and 'Aberdonian', and the Great Western's 'Cornish Riviera Express'.

It was some time after the war and nationalisation that names were restored, but not all were revived and many of the pre-war names simply disappeared.

The 'Capitals Limited' was an uninspiring choice of name for a new King's Cross-to-Edinburgh nonstop in 1949, although calling it the 'Elizabethan' from 1953 was better. Other new names also stood the test of time, even if the trains did not; the East Coast's 'Talisman' and West Coast's 'Caledonian', introduced in 1957, are two of the best-remembered trains of the steam age.

Train naming in Britain is now rather half-hearted, but this simply reflects the consistent standard of speed and service interval now being achieved, with no one train on a route standing apart from the others. In recent times, High-Speed Trains have even had the train name simply pasted on the front. In such circumstances, it is perhaps better to just abandon the name altogether, but no wonder the British public still hankers after real engines with real names hauling real trains distinguished by a cast metal plate pronouncing itself the 'Flying Scotsman'.

The LNER constituent companies were generally not known for naming their engines. The Great Northern, Great Eastern and North Eastern Railways each had only one or two named engines, while the Great Central and the Great North of Scotland Railways had more 'namers' but not by any means all of their express engines.

Only the North British Railway bestowed names on all of its passenger engines, and they were mostly names inspired by the romantic and beautiful country the line ran through, such as glens, Scottish castles or characters from Sir Walter Scott's novels. Even some goods engines carried names, to commemorate their exploits on the continent during the First World War. The NBR, though, never gave its engines cast nameplates; they remained painted on, right through LNER and BR days.

While Gresley of the Great Northern became Chief Mechanical Engineer of the LNER, the first chairman of the new company was

William Whitelaw, a North British man, and he appears to have influenced the new company's decision to start naming its locomotives. First came the five NER Raven A2 Pacifics, given city names connected with the area they worked in. The GNR A1s, likely to be far more numerous, would soon have exhausted this theme, and somehow Welwyn Garden City, Grimsby and Biggleswade may not have had quite the right ring to them.

The first A1 Pacific, No. 1470 was named *Great Northern* by the Great Northern Railway. Sir Frederick Banbury, the last chairman of the GNR, was commemorated on No. 1471, and for the British Empire Exhibition at Wembley in 1924, No. 1472 became No. 4472 *Flying Scotsman*. Next came No. 4473 but, like No. 1472 initially, it remained unnamed. When No. 4472 received its name *Flying Scotsman*, this also happened to sound like a good name for a racehorse, a happy coincidence as the town of Doncaster was famous for two things: horseracing and building steam engines.

Although it is believed there was never an actual racehorse called Flying Scotsman, this seems to have been the inspiration behind the decision to name the rest of the class after racehorses, which was undoubtedly a real public relations masterstroke.

By choosing the names of the winners of classic races, many of them winners of the St Leger run at Doncaster, it cemented the relationship between the railway and horseracing industries of Doncaster, and helped some of the LNER's express engines to become household names.

It should not be forgotten that the clientele who used the long-distance express trains in the 1920s and 1930s were very much the riding and shooting classes, and the racehorse names would strike a chord with many of the passengers.

Naming individual steam engines is not exclusively a British tradition, but it has never been as widespread in other countries. Early American engines were frequently named, but few European engines ever carried names of any description. Some Asian and African countries also bestowed names on engines in the early days of their railways, but only Britain continued to name virtually all of its passenger engines right up to the end of steam and beyond. This had been inconsistent up to 1923, but once the LNER joined the other three Big Four companies, it was to become accepted practice for very many years.

It was the LNER which was ahead of the game here. While the other companies continued the military and patriotic traditions inherited from their constituent companies, the LNER went first for racehorses, then fast-flying birds and antelopes, and even football teams. Many of these names aroused more public interest, were often associated with speed and lent themselves sometimes to elaborate naming ceremonies.

This was to set the scene many years later when BR started to name diesel locomotives after city councils, corporate customers or TV programmes, where they could exploit every last bit of local publicity from a naming ceremony; the name often only being carried on a temporary basis.

Doncaster works had little experience of casting nameplates, having only ever produced the two tiny 'Henry' and 'Oakley' plates for the pioneer Atlantic No. 990. The A1 Pacific plates not only had a very rough finish by comparison with those on other railways, but they were too thin, and tended to crack. Even *Flying Scotsman* lost its name at one point when the plate cracked in two.

New, thicker plates were designed, with large mounting brackets. These are now unpopular with collectors as they are difficult to mount flush on a wall. The curved A1-style nameplates that were carried over the centre driving wheel splashers were all of a standard size, regardless of the number of letters in the name.

LOCOMOTIVE-NAMING POLICY OF BRITAIN'S RAILWAYS PRE-1923

» **Great Western Railway:** all passenger engines named: e.g. cities, stars, earls, saints.
» **South Eastern & Chatham Railway:** engines unnamed in later years.
» **London Brighton & South Coast Railway:** all passenger engines carried painted names, often towns served by the railway.
» **London & South Western Railway:** most engines unnamed.
» **Great Eastern Railway:** only one engine named.
» **Great Northern Railway:** only one engine named before No. 1470 *Great Northern* in 1922.

» **Great Central Railway:** many engines named; directors, military etc.
» **Midland Railway:** engines unnamed.
» **London & North Western Railway:** passenger engines named; patriotic, military, royalty, mythology, astronomy etc.
» **Cambrian Railways:** engines unnamed.
» **North Staffordshire Railway:** engines unnamed.
» **Furness Railway:** engines unnamed.
» **Lancashire & Yorkshire Railway:** engines unnamed.
» **North Eastern Railway:** only one engine named.
» **North British Railway:** all passenger engines named; Sir Walter Scott novels, glens etc.
» **Caledonian Railway:** many engines named; usually Scottish, including directors' houses.
» **Glasgow & South Western Railway:** engines unnamed.
» **Highland Railway:** passenger engines named; usually Scottish e.g. lochs, bens.
» **Great North of Scotland Railway:** a few engines named; various, including military.

POST-GROUPING

» **Great Western Railway:** castles, kings, halls, manors, granges, counties.
» **Southern Railway:** King Arthur legends, schools, military and maritime, West Country places, shipping lines.
» **London Midland & Scottish Railway:** military, royalty, Commonwealth, legendary figures, cities.
» **London & North Eastern Railway:** racehorses, directors, birds, Commonwealth, stately homes, **foxhunts, football teams, antelopes.**
» **British Railways (steam):** famous Britons, Scottish themes, firths, clans

Chapter 7

'THE TON!'

∗∗∗

ALTHOUGH STEAM engines of a particular class might be expected to be identical, in practice they are not, and some engines gained reputations among their footplate crews as good or bad steamers, rough or smooth riding, reliable, fast, or otherwise.

Flying Scotsman was never considered by its crews to be the best of Gresley's A1s, but its name made it the preferred choice for any publicity-oriented jobs on the LNER. It worked the first King's Cross-to-Edinburgh nonstop in 1928 which, at the time, was still timed to take eight hours 15 minutes, a schedule agreed by the East Coast and West Coast companies in 1888, and amazingly still adhered to 40 years later. The challenge to the drivers was not how to break speed records but how to run slowly enough to not run ahead of time and to accomplish the journey nonstop.

In fact, the fastest train in Britain in the early 1930s was the GWR 'Cheltenham Spa Express', known unofficially as the 'Cheltenham Flyer', on which the Castle class 4-6-0 was booked to average 71.3mph between Swindon and Paddington. The GWR was never a railway to push for maximum top speeds though, and had never really publicised the fact that 4-4-0 No. 3440 *City of Truro* had allegedly hit 102.3mph in 1904.

Gresley intended to introduce a new, long-distance, high-speed train service in 1935 and conducted some tests with A1 and A3 Pacifics, during which first *Flying Scotsman* reached 100mph, then *Papyrus* 108mph.

On November 30, 1934, with renowned driver William Sparshatt in charge, A1 No. 4472 took four coaches 185.8 miles from King's Cross to Leeds in 151 minutes six seconds. Another two coaches were added and, between Little Bytham and Essendine, in Lincolnshire, No. 4472 is claimed to have broken the 100mph barrier, with dynamometer car records to prove it.

In March 1935, again with driver Sparshatt at the regulator, A3 No. 2750 with six coaches ran from King's Cross to Newcastle and back. The 500 miles were covered in 423 minutes 23 seconds, including 300 miles at 80mph average.

Despite scientific evidence from the dynamometer car, experts now believe that *Flying Scotsman* probably only touched 98mph, but *Papyrus* certainly reached 108mph. The lessons learned led to the building of the A4 Pacifics, Gresley's streamlined development of the A1/A3.

The basic design of the A4 was similar to the A3 but with a higher boiler pressure, at 250psi, and slightly reduced cylinder diameter, to give greater power. It was the appearance that was radically different: the streamlined casing, inspired by the shape of a Bugatti racing car and perfected in wind-tunnel tests, set off by a silver colour scheme, making the A4 the most striking steam engine ever seen. It is questionable whether streamlining made the engines any faster, but it certainly aroused huge media interest.

The engines were designed to work the 'Silver Jubilee' the 232.3 miles from Darlington to King's Cross at an average of 70.4mph, only slightly slower than the GWR train but more than twice the distance. On the press demonstration run of September 27, 1935, the new Gresley A4 No. 2509 *Silver Link* twice hit 112mph, taking the world record away from the A3 *Papyrus*. This made *Silver Link* a household name at the time, but its fame has not endured to anything like the extent of *Flying Scotsman*.

When the second streamlined train, the 'Coronation' was introduced, it brought the King's Cross-to-Edinburgh time down to six hours, with one stop, and narrowly beat the scheduled nonstop average speed of the 'Cheltenham Flyer'.

The LMS countered with its 'Coronation Scot' streamliner running between Euston and Glasgow. Again, it was the press trip, headed by No. 6220 *Coronation*, that broke the record, reaching 114mph before having to brake (far too late) for Crewe station. In view of how close this train came to disaster, taking 25mph-restricted junctions and crossings at more than 70mph, a new gentlemen's agreement was entered into by the LMS and LNER to stop competing for record top speeds – at least not with trains carrying passengers.

Gresley still wanted to hold the record, though, and arranged some braking tests, on which the engine to be used was the fairly new but well run-in double-chimneyed A4 No. 4468 *Mallard*. Although the tests were to be conducted south of Peterborough, Gresley asked for the train to go north to Barkston Junction, to return down Stoke Bank, the line's fastest stretch of track. The officials on board were told at Grantham they would be trying for the record and were given the option of disembarking. None did, and No. 4468 topped 126mph, setting the world steam speed record that has never been bettered.

By then the LNER's naming policy had moved on and, although *Mallard* and *Silver Link* both became household names in the 1930s, *Silver Link* was probably the second best-known engine after *Flying Scotsman*, by virtue of its silver livery and dramatic public debut rather than by its name. The A4 names just did not capture the public's imagination quite like the racehorses, *Flying Scotsman* especially.

The outbreak of war stopped any further record attempts, and it was well into the BR era before speeds returned to anything like their pre-war levels. It is a Gresley A4 Pacific, No. 60007 *Sir Nigel Gresley*, which holds the post-war British steam speed record – 112mph – set on Stoke Bank in 1959.

In later BR steam days, 100mph-plus speeds were relatively commonplace, although rarely authenticated. Some of Gresley's A4s, now 30 years old, were still achieving the ton on the Glasgow-to-Aberdeen three-hour expresses in the mid-1960s and, in particular, several Southern Railway Bulleid Pacifics are said to have achieved speeds of more than 100mph in their last few days of service on the Bournemouth main line in 1967.

We are a little parochial in Britain, though, and we forget that many other countries built some superb express steam engines which, in most cases, had much longer, straighter, faster stretches of track to run on. Many American engines are claimed to have exceeded 100mph and, although there is insufficient definite proof, it is highly likely that Philadelphia & Reading Railroad 'Camelback' Atlantic No. 343 reached 100mph on June 14, 1907, and Pennsylvania Railroad E6 Atlantic No. 460 did likewise on June 11, 1927. The fastest were the Milwaukee Road 4-6-4s, of which No. 6402 on July 20, 1934 in tests for the introduction of the high-speed 'Hiawatha' service between Chicago and Milwaukee, averaged 90mph for 69 miles, undoubtedly exceeding 100mph in the process.

The 'Hiawatha' in regular service was actually scheduled to average 100mph, and it is known that one held the world speed record before *Mallard*.

No doubt many French and German engines could also claim the unofficial record, but the only scientifically validated feat was 124.5mph by a streamlined Bavarian Pacific in 1935. This engine actually achieved this speed on two separate occasions and on an undulating section of track, whereas *Mallard*'s 126mph was all downhill. Scientific evidence of the German engine's feats exists, but is incomplete. It is highly likely that it may have just touched a higher maximum than *Mallard* but, with the Second World War looming, the rest of the world was not really interested in German claims to world records at the time.

Again, the LNER's marketing 'spin' seems to have worked: *Flying Scotsman* is believed by many to have been the first 100mph locomotive and Gresley's *Mallard* has come to be regarded as the fastest steam engine in the world.

The Second World War, of course, put a stop to speed exploits, but there was no rest for Gresley's Pacifics, including *Flying Scotsman*. In March 1939, the engine was transferred from King's Cross shed back to its original home depot of Doncaster and, as late as 1943, it received a coat of all-over wartime black paint. Express train haulage was at an end for the duration but, like all steam engines during the war, No. 4472 was set to work hauling heavy troop trains and even munitions and coal trains, and not necessarily on its normal operating routes.

Railway operating in the war years is not particularly well documented, but there were certainly instances when the most unlikely engines found their way to the opposite end of the country from where they were normally to be seen. In *Flying Scotsman*'s case, it changed depots in 1944 more times than most LNER Pacifics did in their entire working lives, moving first from Doncaster to New England at Peterborough, then over to the one-time Great Central shed at Gorton, in Manchester, briefly back to King's Cross, but then New England again, and then back to Doncaster.

Sir Nigel Gresley died in office in 1941, and was succeeded by Edward Thompson, a man with a very different engineering background, who later achieved some notoriety by completely rebuilding Gresley's pioneer A1 Pacific No. 4470 *Great Northern*. Thompson did not just rebuild engines, he undermined all of Sir Nigel Gresley's design principles, such as building steel coaches instead of teak ones.

There were issues that needed addressing, and one was the chaos that the LNER's locomotive numbers had been allowed to descend into. It had made sense when the various independent companies had been combined in 1923 and the LNER had found itself with many engines that duplicated each other's numbers, but then new engines had been given numbers just to fill gaps in the sequence and, by 1946, a new approach was required. The original system had not really taken account of the huge numbers of engines Gresley was going to build.

Thompson had two attempts to make order out of chaos and, as a result, on January 20, 1946, No. 4472 *Flying Scotsman* became No. 502, but by May 5 it had changed again, to No. 103. It was just a short time later, on November 18, 1946, that *Flying Scotsman* entered Doncaster works for rebuilding to A3 specification.

On January 4, 1947 No. 103 returned to service in post-war Thompson apple green livery. This was little different to pre-war style except that the ornate shaded gold-leaf lettering was replaced by a slightly more austere plain yellow, with a slight reduction in lining-out, for example on the back of tenders.

Chapter 8

LNER PACIFICS

✳✳✳

G RESLEY'S GREAT Northern Railway Pacifics were referred to as the A1 class by the GNR and subsequently the LNER and, logically, the five Pacifics built by Sir Vincent Raven for the North Eastern Railway were referred to as the A2 class. When the A1 design was upgraded, the new engines took the classification A3 as they were quite different, despite being little changed externally.

Next on the scene were Gresley's streamlined Pacifics in 1935 and these took the classification A4. So far it was all quite logical. However, as new A3s were being built, older A1s were being rebuilt to A3 specification so, as they were reclassified; the A1 class was expected to be phased out. By now the classification A5 had been allocated to the Pacific tank engines inherited from the Great Central Railway, with other 4-6-2Ts taking other A-class numbers.

When Gresley died in 1941 and Edward Thompson took over as the LNER's Chief Mechanical Engineer, he introduced some new designs, but Gresley designs continued to be built. Thompson disagreed with many of Gresley's design principles. He took Gresley's P2 2-8-2s and rebuilt them as Pacifics in 1943, calling them A2s, as by 1937 the five NER-built A2 Pacifics had been withdrawn and scrapped, so the classification A2 was spare.

The LNER was still turning out Gresley's V2 class of 2-6-2, which had acquitted itself brilliantly in wartime service, its smaller wheels giving the power to move prodigious tonnages. Thompson had the last four completed as Pacifics in 1944 and also called them A2s although, apart from the size of their driving wheels, there was little discernible similarity between the two new designs of A2.

Gresley had been a firm believer in standardisation of parts, so A1/A3s and A4s had identical 6ft 8in diameter driving wheels, while V2s and P2s had identical 6ft 2in diameter driving wheels. Although complicated enough, this is by no means the end of the story of the LNER Pacific classes, as Thompson felt he could even improve on Gresley's Pacific design.

He went further than just building a new one; he actually rebuilt Gresley's original masterpiece, none other than the pioneer A1, No. 4470 *Great Northern*, as his prototype. Rightly or wrongly, it was felt that the new design should be called the A1 despite the fact that there were still many of the original A1s still running.

Wartime had held up the rebuilding programme and delayed production of Thompson's 'new' Pacific until the end of the war. However, in April 1945, the last unrebuilt A1s which, perhaps surprisingly, included *Flying Scotsman*, were reclassified A10. Thompson's A1 still had three cylinders, albeit with three separate sets of valve gear, and there were some external design similarities to Gresley's non-streamlined Pacifics, but it was a more functional, less complicated design built for power and ease of maintenance, not speed and elegance. *Great Northern* kept its name though, and further engines in the class mostly continued the tradition of being named after racehorses.

One of Thompson's design principles, which he did not share with Gresley, was that all connecting rods, i.e. those from the pistons to the driving wheels, should be of equal length; Gresley was happy to vary the length provided they all drove on the same axle. Thompson's three-cylinder Pacifics therefore looked ungainly, with the outside cylinders set well back to drive the middle axle, while the inside cylinder was well forward between the frames but drove the leading axle using the same length connecting rod as outside. It was perhaps fortunate that the war ended and Thompson could build new engines, or the remaining

17 original A1s could easily have been rebuilt to Thompson A1 design, and the original *Flying Scotsman* could have disappeared in the 1940s.

As it was, the pre-war rebuilding programme was resumed, with *Flying Scotsman* finally becoming an A3 in 1947 and the last one, No. 60068 *Sir Visto*, in BR days in 1949. It was left to A.H. Peppercorn, who briefly succeeded Thompson just before nationalisation, to build some new A2 Pacifics, to a slightly more conventional external style, and the various Pacifics with 6ft 2in driving wheels were known as A2s, A2/1s, A2/2s and A2/3s. Peppercorn also started building the rest of the A1 class started by Thompson, but most of the Peppercorn A1 and A2 Pacifics were actually built by British Railways after nationalisation.

They were good, strong, reliable and, above all, economical engines; they just lacked something of the style of *Flying Scotsman* and the Gresley Pacifics, but they had been built in totally different circumstances and, in post-war operating conditions, the various LNER Pacifics all ran very successfully side by side.

All of Peppercorn's A1s were scrapped by 1966, but the last working A2, No. 60532 *Blue Peter*, was purchased by the late Geoff Drury in 1969 and has seen main line action in recent years. The A1 Locomotive Trust, however, constructed a brand-new Peppercorn A1; it first steamed in 2008, was known as No. 60163 *Tornado*, and quickly established itself as Britain's best-known express steam locomotive. It will be fascinating to see whether *Flying Scotsman* can retake that accolade from the newcomer now that the A3 has finally returned to steam. From the public interest shown in the first few weeks of its return to action, it would certainly seem that the British public has lost none of its affection for *Flying Scotsman*.

LNER PACIFIC TENDERS

Gresley's A1s had been built with eight-wheel tenders in GNR style with coal rails. In fact, these were a little too big at first and turntables had to be extended to accommodate them. *Flying Scotsman* even had to borrow a six-wheel tender for the second Wembley Exhibition in 1925, as it was too long with its normal eight-wheeler.

For the nonstop runs to Newcastle and Edinburgh, new tenders were built with corridors along the side, easily distinguishable as they were much taller, without coal rails. The A1s and A3s with corridor tenders kept them only until new A4s took over all the nonstop runs in 1936. However, new corridor tenders were built for some of the A4s, which were streamlined themselves, and some A4s had new streamlined non-corridor tenders. The A1s and A3s that lost their corridor tenders received new, non-streamlined but high-sided ones, and there was an element of tender-swapping after this, and a small number of A3s could always be seen with high-sided, but non-corridor tenders. *Flying Scotsman* had a corridor tender from 1927 but when it lost it to an A4 in 1935, it received a high-side one, so the refitting of a corridor tender in 1963 after preservation did not, in itself, dramatically alter the engine's appearance.

Chapter 9

BRITISH RAILWAYS

***

HE SECOND World War took its toll on Britain's railway system. Some companies, particularly the LNER, were in financial difficulties throughout their existence, and the decision was soon taken by the post-war Labour Government to nationalise the railway system. The much-loved British Railways came into being on January 1, 1948.

It must have been an incredible upheaval for everyone concerned but for the railway enthusiasts at the time, it was a fascinating few years. BR of course, inherited thousands of steam engines, many of which should have been withdrawn years earlier, and many of which carried the same number as at least two other engines. There were far bigger priorities at the time, but steam enthusiasts were interested in the exchange trials, the new numbering system and the new liveries.

BR was organised into six regions. The Eastern Region was made up of the former GNR, OCR and GER routes of the LNER. The North Eastern Region comprised the ex-NER routes, and the Scottish Region combined all LMS and LNER routes north of the border. *Flying Scotsman* was an Eastern Region engine, and acquired the prefix E to its number 103 for a while in March 1948, though retaining LNER apple green livery, but with 'BRITISH RAILWAYS' painted on its tender.

To make decisions on future locomotive policy, engines from the Big Four companies were tested against each other on each other's routes. This did not involve *Flying Scotsman* but, interestingly, it did involve *Mallard*. The Eastern Region chose Gresley A4s, rather than the newer Peppercorn A1 Pacific, to prove the superiority of the engines it had inherited. *Mallard* and the other A4s proved strong, fast and economical, but not reliable, as the middle big-ends played up more than once on test.

As Robert Riddles, an LMS man from Crewe, was appointed BR's first Chief Mechanical Engineer, LMS design policy inevitably took the lead after 1948, and three-cylinder designs, with conjugated valve gear, were not going to be the way forward.

BR wanted to establish a corporate identity totally different to its predecessors and, after some experiments, including some particularly lurid colours, settled on blue for express engines, LNWR-style lined black for mixed-traffic engines and plain black for goods engines. GWR Brunswick green was adopted for passenger engines but, in practice, not many classes acquired this livery at first.

No. E103 *Flying Scotsman* was rightly nominated for express blue and emerged in this colour on December 16, 1949, with black and white lining, cream numbers and, for the first time, a smokebox numberplate, a tradition adopted from the LMS. By now a numbering system had been devised and 60,000 was added to all ex-LNER engine numbers, so *Flying Scotsman* was now No. 60103. A badge was designed for the tenders, often referred to as the cycling lion, and this replaced the words British Railways. Perhaps the major cosmetic difference, apart from the colour, was that the wheels and frames were black and unlined.

✳ ✳ ✳

When built, Gresley's A1s were considered almost too heavy even for the Great Northern main line, and some modifications had taken place before they could venture to Scotland. With new A1 and A2 Pacifics being built after the war, and considerable numbers of the smaller V2 2-6-2s now available, the LNER, and later BR, gave thought as to whether Pacifics could be used to accelerate services on other lines. The Great Central route from Marylebone to Sheffield and Manchester had already seen Pacifics; the Great Eastern main line from Liverpool Street to Norwich, still worked by underpowered 4-6-0s, was another possibility.

The GCR, having been built late, and to a generous loading gauge with Channel Tunnel freight traffic in mind, presented no great problem and in 1944, *Flying Scotsman* had a brief spell at Gorton shed in Manchester, from where it worked from London Road station over the Woodhead route over the Pennines to Sheffield. On June 4, 1950, No. 60103 was transferred back to the GC line, allocated to Leicester shed, from where it worked over the main line out of Marylebone. The use of A3s (and V2 2-6-2s) on the Great Central was to last only until a major regional reorganisation in 1958, when the line became part of the London Midland Region. For the first time, a boundary could be drawn between the regions, instead of following the pre-nationalisation pattern, which had considerable overlapping of routes.

It was during its brief fling on the GCR that *Flying Scotsman* lost its BR blue livery. It was found that the colour could not be touched up after repairs, and a total repaint was necessary each time. On March 14, 1952, No. 60103 emerged from Doncaster works in Brunswick green, which it was to carry for the rest of its BR career. Although basically a Great Western livery, the black and orange lining was in a different style from the GWR. The tender badge and cream numbers were retained and, in common with normal Doncaster practice in BR days, the cylinder covers were unlined black.

On November 15, 1953, *Flying Scotsman* returned to the GN main line but was allocated to Grantham, which was the normal engine changing point for East Coast Main Line expresses. It had a brief spell at King's Cross in 1954 and from April 7, 1957 found itself permanently back at 'Top Shed', reallocated to King's Cross (34A). LNER engines traditionally spent long periods allocated to one shed, and some were never reallocated in their entire lives, so *Flying Scotsman* was actually relatively well travelled.

It was not until well after the end of the war, and nationalisation in 1948, that the LNER Pacifics returned to their former glories. The A3s though, were overshadowed by their A4 successors, which monopolised all the top jobs. *Flying Scotsman*, the engine, rarely if ever, hauled the

'Flying Scotsman' train in BR days. During the post-war period, *Flying Scotsman* was just another of BR's A3s and did very little of any note.

In fact, with better track, better signalling and less traffic, the pre-war racehorses could probably have improved on their speed exploits but economics and efficiency were more important to BR. The element of competition had gone and BR had no interest in proving that its older engines were faster than its own new designs. Before long the future was going to be diesel anyway.

Perhaps the most positive move, though, was the long-overdue fitting of a double chimney to *Flying Scotsman* in 1958, which transformed its performance at the ripe old age of almost thirty-five. In LNER days, Gresley carried out numerous experiments with his A1 and A3 class, one of which was the fitting of a double chimney and Kylchap double blastpipe to No 2751 *Humorist* in 1937. These had first been tried on the later A4s, *Mallard* having been built with a double chimney, and the earlier engines were eventually converted.

THE KYLCHAP

The 'Kylchap' exhaust system originated with the Finnish engineer Kylala, and was perfected by André Chapelon, the premier French locomotive designer. Gresley consulted Chapelon regarding the fitting of the blast-pipes to his locomotives and what the optimum dimensions would be.

The blastpipe is a hole above the cylinders which allows the steam to escape through the smokebox to the chimney. If the hole is too small, the steam cannot escape quickly enough, so back pressure builds up and slows the engine down. The precise size of the hole is critical to a steam engine's performance, and any double blastpipe design will assist by doubling the speed at which steam can escape. The Kylchap system is much more complicated, though, with a series of cowls fitted between the blastpipe top and the chimney.

The last four A4 streamliners, including *Mallard*, were built with double chimneys and Kylchap double blastpipes, after experiments on an A3. But, despite the obvious improvements in performance and economy, these were not adopted as standard by the LNER.

Appropriately enough, the engine selected for most of the experiments was No. 2751 *Humorist*, and some of the smokebox modifications were more comical than effective. These experiments had actually started with the engine still in single-chimney form but, when all the experiments failed, the double chimney and Kylchap were fitted and these were more successful. One drawback was a very soft exhaust, totally different to the loud bark of a single-chimneyed Gresley engine and the drifting smoke from which seriously affect the driver's forward visibility.

It was not until the late 1950s, after numerous experiments and negotiations between locomotive depots, particularly King's Cross, and the main works at Doncaster, that a decision was made to convert the rest of the A4s to the double-chimney arrangement. The cost was £200 per engine and, apart from the much more free-running engine, there was a saving in coal of 7lb per mile. Perhaps surprisingly, the rebuilding programme immediately continued with the A3s being treated at a cost of £153 per engine for a saving of 6lb per mile.

Flying Scotsman received its double chimney in December 1958. Drifting smoke was still a problem but BR was so impressed with the economies obtained that it seems surprising it was so slow in addressing this issue. One, No. 98 *Humorist*, the unique double-chimneyed engine for many years, had been fitted with smoke deflectors in A1/A2 style by the LNER in 1947, but little is known of why they were not later adopted widely. The eventual solution is probably one of the most bizarre chapters in the story of British steam traction.

There had been some further half-hearted experiments with tiny deflectors alongside the chimneys but with little success, but it was Peter Townend, shedmaster at King's Cross, who suggested the fitting of smoke deflectors in the style adopted widely by the Deutsches Bundesbahn, which he had seen in Germany. A photograph of a German Pacific was sent to Doncaster works, some deflectors were designed and manufactured and tried out on an A3, and they worked. Not only did Doncaster start fitting them as A3s passed through works, but a further batch was manufactured and sent to King's Cross to be fitted at the depot.

It was late in the day, though, and not all A3s even received double chimneys, and of those that did, several never actually acquired the smoke deflectors. *Flying Scotsman* was fitted with its smoke deflectors in December 1961, and carried them for just over 12 months.

There is no doubt though, that the double chimney and Kylchap exhaust system, transformed the performance and economy of both the A4s and A3s and although the streamlined Pacifics simply continued much as before, the A3s, *Flying Scotsman* included, which had been relegated to the slightly less arduous main line jobs for many years, found themselves back on 'top-link' duties again, for the last year or two of their 40-year careers.

In fact, operations changed quite drastically towards the end of steam, and the traditional engine changes were largely abolished. A3s and A4s started to work throughout between King's Cross and Newcastle on a far more regular basis, something which the A3s had not done since the A4s had taken their top jobs in 1935.

THE END OF THE A3 PACIFICS

In 1955, BR announced its modernisation plan, a major part of which was the elimination of steam traction. In the post-war era, trains on Britain's fastest main line were still being hauled by steam engines designed more than 30 years earlier, soon after the end of the First World War. *Flying Scotsman* and the A3 class may have been overshadowed by the streamlined A4s on the top duties, but the later Peppercorn Pacifics had merely assisted them, never replaced them, and the even-newer BR Standard designs, such as the Britannia Pacifics, were no competition at all.

Diesels, though, were a threat. Even Gresley had looked into converting to diesel and had visited Germany to travel on the 'Flying Hamburger' diesel express before the war, only to conclude that steam could do anything diesel could do. Britain was, in fact, rather short-sighted in this respect. The US was largely dieselised early in the 1950s and most European countries used modem traction quite widely well before Britain. Even after the 1955 announcement, it would be nearly ten years before steam's supremacy could really be said to have ended in Britain.

In *Flying Scotsman's* case, the first competition came when a few English Electric Type 4 2000hp ICo-Col diesel electrics were introduced on the East Coast Main Line in 1958, but they were no more powerful, and perhaps rather less reliable. The English Electric Deltic Type 5 3300hp Co-Cos were a more serious matter. Only 22 were built as they were phenomenally expensive, but they were able to replace all the A4 streamliners, so it only needed another class of powerful and reliable diesels to arrive in quantity, and that would spell the end for all the LNER Pacifics.

That class was the Brush/Sulzer Type 4 2750hp Co-Co, the first of which, D1500, started work on the ECML in 1961. The naming of the King's Cross Deltics after Doncaster race winners was a good way of continuing the tradition that was effectively started by *Flying Scotsman*, but Crepello and Ballymoss would never become household names like their steam predecessors. The public was far more interested in fast cars and aeroplanes now, not railway engines.

On December 7, 1959, No. 60104 *Solario* became the first A3 Pacific to be withdrawn from service to be scrapped, from King's Cross shed, after suffering accident damage, but it was to be another two years before the second one was withdrawn, this time No. 60085 *Flamingo*, from Carlisle (Canal) shed. Several A3s were allocated here for working the ex-North British Waverley route to Edinburgh.

In later years, a few A3s even found themselves working over the one-time LMS Settle & Carlisle route between Leeds and the border city, one of the few instances where the class was ever seen away from the traditional LNER routes they were designed for. From then on, withdrawals accelerated and, on January 14, 1963, No. 60103 itself was withdrawn from King's Cross shed after running 2,076,000 miles. On that day, *Flying Scotsman*'s last run for BR was on the 1.15pm from King's Cross to Leeds as far as Doncaster, attracting considerable media interest.

By now, dieselisation was taking effect rapidly and, in May 1963, King's Cross shed closed to steam completely, with a virtual ban on steam enforced south of Peterborough. It proved impossible to stop occasional steam workings because of diesel failures, and in 1964, A3 No. 60106 *Flying Fox*, BR's oldest express steam engine, worked what proved to be the last steam-hauled up 'Flying Scotsman' into King's Cross, after the failure of a Deltic diesel, losing only two minutes on the schedule.

The remaining A3s in England in 1964 were relegated largely to standby duties, particularly at Gateshead and Darlington, with a handful at New England shed, Peterborough that worked occasional parcels, freight and relief passenger trains. The NER ones were withdrawn by November 1964 and on December 26, 1964, the last three English A3s were withdrawn from New England and Nos 60062 *Minoru*, 60106 *Flying Fox* and 60112 *St Simon* went for scrap.

They had their final fling on railtour duties, with No. 60112 hauling a train from Waterloo in August 1963, clocking 90mph on the South Western main line with just six coaches. Three A3s soldiered on in Scotland in 1965 though, allocated to St Margaret's shed and mainly working from Edinburgh over the Waverley route, usually on goods trains.

No. 60100 *Spearmint* was withdrawn in June, No. 60041 *Salmon Trout* in December and on January 17, 1966, the last A3 Pacific in BR service was withdrawn without ceremony: No. 60052 *Prince Palatine*. Its last railtour duty had been a run from Edinburgh to Aberdeen on September 4, 1965. It was not the last to be scrapped, but when *Salmon Trout* was broken up in September 1966, this left just one of the 78 A3s still in existence.

Gresley's A3 Pacifics did not disappear from the East Coast Main Line though. Even by the time the last English A3s were withdrawn from regular service, *Flying Scotsman*, already the most famous of them all, which had been withdrawn nearly two years earlier, could occasionally still be seen, though not at King's Cross, and it presented a rather different appearance than most people were used to by then.

Chapter 10

ALAN PEGLER: *SCOTSMAN'S* SAVIOUR

✳✳✳

THERE COULD very easily have been no A3 Pacifics left at all, if it had not been for Nottinghamshire businessman Alan Pegler, the first of several larger-than-life characters to have become involved with *Flying Scotsman* after its withdrawal from service.

Alan had seen it at the British Empire Exhibition at Wembley in 1924, and it had left a lasting impression on the young lad, such that, 40 years later, when he felt able to save the engine from being scrapped, he went ahead and bought it from British Railways for £3,000. He perhaps had a slight advantage, having served as a part-time member of the British Railways Board.

Pegler was not new to steam preservation nor to organising special trains on BR: he had been behind the operation of the two preserved Great Northern Atlantics from York Museum on specials from King's Cross in 1953. He had also been more than just a founder member and prime mover in the campaign to reopen the Ffestiniog narrow gauge railway in North Wales which, from very small beginnings in 1955, had already grown into a major tourist attraction by 1963, and was to progress to even greater heights. Alan actually bought the railway himself, vested it in a trust to avoid legal complications, and remained very much involved until his death in 2012.

Part of Pegler's deal with BR was that *Flying Scotsman* would be returned to LNER single-chimney form at Doncaster works, though still as an A3, and repainted into LNER apple green livery as No. 4472. In fact, at this time it is doubtful if BR would have allowed it to remain in BR livery in private hands. Alan not only arranged for the engine to be allowed to haul special trains on the national network, which was not an entirely new concept, but he entered into a legally binding written agreement with BR for a term of three years, later extended to eight years, a unique contract that no other engine owner was ever privileged to obtain. The deal did not quite extend to his driving his own engine on these tours, as the unions felt that this could threaten their members' jobs.

With Doncaster works still overhauling steam engines on a daily basis, the job did not take long, and a partially unpainted single-chimney A3 was given a couple of test runs to Peterborough before entering the paint shop for the final touches. The engine was certainly not returned to original condition, but the single chimney enabled the smoke deflectors to be removed and it certainly looked very different to the engine that had hauled the 1.15pm King's Cross to Leeds on January 14. A major alteration that did not radically alter its appearance was the new tender, a corridor-fitted one transferred from the recently withdrawn A4 No. 60034 *Lord Faringdon*, which no longer needed it, but which would no doubt be useful in the future.

On March 26, 1963, No. 4472 *Flying Scotsman* emerged from the works, looking as good as new, and on April 16 Alan Pegler formally took possession of what was to become his pride and joy. A noticeable change from normal LNER practice was that the nameplates were painted red.

With steam traction still in everyday use in most parts of the country, *Flying Scotsman* ran more or less like any other steam engine operated by BR, although its runs tended to be longer than most steam duties were by that time. It certainly reached parts of the country it had never seen before but, as steam was rapidly phased out, crucial infrastructure was equally quickly removed, particularly water columns and water troughs. About 80 miles is the normal prudent limit with one tender full of water, and a quick five-minute top-up from a column is then the normal practice, as messing about with fire hydrants, pumps, road tankers and long lengths of hosepipe is time-consuming and inconvenient.

The first public outing for the preserved No. 4472 was from Paddington to Ruabon on a special bound for the Ffestiniog Railway. *Flying Scotsman* was already famous for its exploits 40 years earlier; now it was going to be seen in parts of the country never before visited by a Gresley Pacific. The A3 found its way to Wales, to Scotland, and to the South Coast. One of the Welsh visits was one of the most notable, on November 13, 1965, when No. 4472 reached Cardiff from Paddington in 2 hours 17 minutes, a record for steam traction.

In late 1964, the engine had an overhaul at Darlington works, which applied its trademark-green cylinder covers. Doncaster had a tradition of always painting cylinder covers plain black. The eventual solution to the water problem was the acquisition of a second tender, which was actually another corridor one from another newly withdrawn A4, in this case No. 60009 *Union of South Africa*, which was also subsequently preserved and acquired a replacement tender.

In September 1966, *Flying Scotsman* emerged from works with two tenders, its first outing in this guise being from Lincoln to Blackpool, a route where the second tender was not really necessary. The second tender carried the number 4472, and the cabsides now carried the LNER coat of arms, just as in the Wembley Exhibition of 1924.

A few weeks later, No. 4472 was able to return to King's Cross for the first time in two years. Without doubt, the most inconvenient removal of watering facilities was at Peterborough, and without either the platform-end columns or the troughs at Werrington just to the north, there was simply no water to be had at just the point it was needed, 76 miles out of the 'Cross'.

Flying Scotsman worked hard for its living, and proved very reliable. June 1967 saw the 40th anniversary of the first King's Cross-to-Newcastle nonstop express, a train that was not in fact, hauled by *Flying Scotsman* in 1927. However, it was an anniversary that ought to be marked by a re-run.

The second tender now gave the engine a range of more than 200 miles without stopping, but nowhere near enough to get to Newcastle. It tends to be forgotten now that although diesels took over from steam, they still used steam to heat their trains, and diesel classes were actually fitted with water scoops so they could top up their train heating boiler water tanks from water troughs. A select few troughs had remained in use to avoid diesels having to make unnecessary station stops to take water, and the ones located at Scrooby, Wiske Moor and Lucker remained in use on the ECML.

No. 4472 did not make Newcastle nonstop this time but it set the scene for the big one the following year.

Circumstances changed in the meantime. By 1967 *Flying Scotsman* had been joined on the main line by other privately owned engines and, in the autumn, the East Coast Main Line saw eight weekends of steam, featuring not just No. 4472, but its streamlined partner, A4 No. 4498 *Sir Nigel Gresley*, and a GWR rival, 4-6-0 No. 7029 *Clun Castle*. Tours ran steam-hauled from King's Cross to Newcastle or Carlisle and return, using one of the three locomotives in each direction, but *Flying Scotsman* was the star. With its two tenders, it was the only one able to handle the King's Cross-to-Peterborough leg of the longer tours, and both the other engines were restricted to running north of Peterborough only.

Clun Castle did make it to King's Cross twice, but only on shorter runs to York, and it had to leave its train at Peterborough for an hour each time and run two miles to New England for water. Even so, soon after the end of regular steam, there were already areas where the operation of privately preserved steam was proving quite difficult.

It was hardly surprising when, at the end of October 1967, BR issued a statement that it was not prepared to continue to operate privately owned steam locomotives on any of its routes — *Flying Scotsman* excepted, of course — Alan's written contract still had four years to run.

But it was not on a totally steamless British Railways that *Flying Scotsman* had what many consider to have been its finest hour. BR's last steam engines were still soldiering on in the north-west of England, although their area of operation was contracting and, after August 11, 1968, there would be no more. It was a few months before this that No. 4472 attempted the impossible.

May 1, 1968 was the 40th anniversary of the first King's Cross-to-Edinburgh nonstop service, a train that *Flying Scotsman* itself had hauled at the tender age of five. Could it be done again? Pegler must have had some very good friends in some very high places to have even dared to suggest it. But, sure enough, at 10am on May 1, No. 4472 set off from King's Cross as it had done countless times before, this time alongside the Deltic-hauled 'Flying Scotsman', and reached Edinburgh 7 hours 45 minutes later, without having stopped en route.

The train had miraculously been allowed to crawl over a broken rail near Doncaster. *Scotsman* had picked up far too little water from water troughs which were only half-full, and the signalman at Berwick-on-Tweed had sent the train into the loop where a road tanker full of cool refreshing water was waiting. The brave decision was made to press on though; the signalman got the message, allowing the train through the loop and back on the main line without the wheels having stopped turning – just. Expert driving saw the train coast the last few miles into Edinburgh Waverley and to a hero's welcome, with barely a drop of water left in either tender. And, if that wasn't enough, the exercise was repeated southbound three days later.

It could be said that any engine with two tenders could have achieved this, and the real achievement was that of the many people involved who made it possible. But it was the fact that *Flying Scotsman* became famous for the achievement in 1928, that was a huge media event itself, an event in which the engine's instantly recognisable name played such a major part, that made the 1968 attempt an even bigger media show.

The BBC was on board and the 60-minute programme that resulted was shown on TV every Christmas for years afterwards.

Little could follow the Edinburgh nonstop, of course. *Flying Scotsman* visited the north-west to play a small part in the run-up to steam's final curtain in the summer of 1968, although it perhaps wisely stayed away for the very end in August.

It did, though, achieve one last honour, being the last steam engine to run over the rival West Coast Main Line route over Shap summit in June 1969. The years from 1963 to 1969 should probably be regarded as *Flying Scotsman*'s golden years as a preserved locomotive, as Alan Pegler continued to run packed trains over all parts of the British Railways system.

But, as much of this period was a time when BR itself was still operating steam, albeit in ever-decreasing numbers, even No. 4472 is now widely remembered as having been just a privately owned steam engine hauling railtours back in steam days. In reality, though, it was much more than this. It had been a legend for 40 years while in everyday service; now it was just starting to do it all over again — and much more was to come!

✻ ✻ ✻

It is often argued that *Flying Scotsman* should have been officially preserved by British Railways when it was withdrawn from service, and that its acquisition by the National Railway Museum in 2004 was 41 years too late.

Locomotives have been preserved both officially and by private individuals or societies since some of the early locomotives, such as *Locomotion* and *Rocket*, were presented by their owners to local museums in the 19th century.

Preservation remained largely official though, i.e. by the railway companies themselves, with particularly historic engines simply put to one side at Doncaster or Crewe works, for example, rather than being scrapped, although unfortunately several were later scrapped, particularly during wartime. Although a small number of engines were put on display, such as the Furness Railway 0-4-0 No. 3 'Old Coppernob' at Barrow-in-Furness in 1900, the first railway museum to be opened was that at York, by the LNER, following the Stockton & Darlington centenary parade in 1925.

Flying Scotsman missed that event and another A1 Pacific, No. 2555 attended, being named *Centenary* in commemoration, a rare exception to the racehorse theme for the class. Private preservation really started when the Stephenson Locomotive Society bought ex-London Brighton & South Coast Railway 0-4-2 No. 214 *Gladstone* in 1927, but there was no intention of running it, and the old engine went into the museum at York.

It was not until BR days that another railway museum opened, at Swindon in 1962, after preservation of locomotives had begun to gather pace after nationalisation. However, the Museum of British Transport had been opened in a former bus garage at Clapham in 1960 and although road transport was also represented, railway exhibits, particularly loco-motives, formed the lion's share of what was on show. Private preser-vation also took off after the Second World War, with first the narrow gauge Talyllyn, then the Ffestiniog Railway being reopened in Wales.

The first standard gauge steam railways to reopen were the Bluebell Railway in Sussex and the Middleton Railway in Leeds, both in 1960, but a year earlier, private preservation had already made a major leap forward. Two standard gauge tank engines were purchased privately from BR and one of them, a Great Northern Railway J52 0-6-0ST, was set to work hauling occasional main line railtours, setting a precedent that larger engines might follow.

BR had found itself with quite a large collection of preserved steam engines, some restored and some not, and some on display and some not. As a result of BR's modernisation plan, and the imminent end of steam, a list was published by the British Transport Commission in 1961 of existing preserved engines, and ones that were to be preserved, to ensure that the widest possible cross-section of British steam locomotive designs could be seen by future generations.

The list was compiled quite scientifically so that, as far as possible, there would be an example of each pre-Grouping and pre-nationali-sation railway, at least one from each of the major Chief Mechanical Engineers, and one of each different wheel arrangement. In fact, it was too late in some cases as, for example, no Cambrian Railways engine remained in existence, nor any 4-4-4 tanks. Nevertheless, there were avoidable omissions, although relatively minor ones.

If there is any criticism with the benefit of hindsight, it is the fact that it was too scientific, and little regard was paid to fame and fortune, or public appeal. For example, a Gresley streamlined A4 Pacific was quite rightly included on the list, but *Flying Scotsman* was not: it was consid-ered to be just another Gresley 4-6-2 and there was no justification in preserving two of them, however famous the A3 might be.

The fact that it was an A3 was in any case a problem in itself. There was a policy decision taken not to preserve engines that were no longer in substantially original condition. In fact, the choice of *Mallard*, the fastest A4, as opposed to *Silver Link*, the first, once the fastest, and arguably the best-known class member, was one of the few exceptions to the scientific rule.

Sure enough, though, something larger than an 0-6-0 tank engine, was soon purchased privately for main line use, and it was a Gresley engine, but it was not *Flying Scotsman*.

K4 2-6-0 No. 3442 *The Great Marquess* became the first main line tender engine to be purchased by a private individual in 1962 but it never quite had the appeal of *Flying Scotsman*, and was not really suited to running long-distance express trains anyway.

The Great Marquess, purchased by Lord Garnock, did run on main line tours in the mid-1960s, and again many years later, when based on the Severn Valley Railway, including a return to its home territory, the West Highland line in Scotland.

The National Collection could never preserve every type of steam engine, but its nevertheless impressive collection has been supplemented by a remarkable variety of privately-preserved engines.

After years of debate, it was eventually decided that the museums at Clapham and York would be closed and replaced by a new National Railway Museum, for which the chosen location was York. It was already an established tourist destination and a great railway centre.

The new museum was opened on September 27, 1975, in the converted one-time North Eastern Railway locomotive shed north of the station, with many of the locomotive exhibits arranged around the two turntables. It quickly established itself as one of the foremost railway museums in the world and has gone from strength to strength.

The NRM involved itself in active steam preservation and many of its locomotives have been seen in steam, both on heritage lines and on main line railtours in all parts of the country. The building had to be rebuilt in the 1980s, but the museum has steadily expanded and has ambitious plans for the future.

✳✳✳

If *Flying Scotsman's* fame and popularity had earned it a place in the National Collection, would the story have been the same? The answer is undoubtedly no. The National Collection expanded rapidly towards the end of steam, but there was no money to restore all the engines, certainly not to steam. Some officially preserved engines did steam in the late 1950s and early 1960s but by 1963, there was no interest in restoring any more, and BR wanted to get rid of all steam as quickly as possible.

The museum at Clapham had been opened in 1960, and space was left for *Mallard*, but this was one of very few engines cosmetically restored in the first part of the 1960s. Only when a regional museum expressed an interest was an engine restored, as happened with LNER V2 2-6-2 No. 4771 *Green Arrow* in 1962, when there was a real prospect of a museum being opened at Doncaster. Had *Flying Scotsman* been available, it would very probably have been chosen instead, and could even have been a static exhibit at Doncaster for the past 50 years.

The Doncaster museum, like most others, never materialised, so *Flying Scotsman*, even if it had been externally restored like *Green Arrow*, might have gone into storage. It could possibly have been selected for another museum that failed to materialise, at Leicester, and perhaps it might have been lent eventually to a railway or steam centre to be returned to steam. Even so, it is doubtful if it would initially have returned to single-chimney form.

Like *Green Arrow*, it may well have appeared in steam at the Shildon cavalcade in 1975, and would no doubt have become a permanent exhibit in the new National Railway Museum at York, from where it might have made railtour appearances. None of this would have added as much to its fame or popularity with the public though; its status would be similar to *Green Arrow* or *Duchess of Hamilton* — still a great and popular engine, but not the most famous locomotive in the world.

✳ ✳ ✳

Alan Pegler's ownership of *Flying Scotsman* may have ended with his bankruptcy in 1972 but he continued to take a close interest in "the old girl", as he called it. Alan's involvement in railway preservation did not start with *Flying Scotsman* of course, as he had already saved the Ffestiniog Railway back in 1955. Even in 2005, he was still that railway's president and he took a short break from working on Ffestiniog business when Brian Sharpe called to see him.

The question that had to be asked was why did he buy *Flying Scotsman?* "It was the colour really." As a very young lad, Alan had been taken to the British Empire Exhibition at Wembley, where he had seen *Flying Scotsman.* What had left an impression on him was the big, highly polished, bright green engine, compared with the relatively drab, dark Brunswick green Great Western Castle that stood next to it.

Many years later Pegler became involved in the family business, the Northern Rubber Company, at Retford and as a result, had become a part-time member of the Eastern Area of the British Transport Commission, as it was the practice for a couple of local industrialists to be included on the board.

In 1963, when *Flying Scotsman* was withdrawn from service by BR, Pegler felt strongly that a non-streamlined Gresley Pacific ought to be preserved. The BTC would not be swayed, though; a streamlined Pacific and a V2 2-6-2 were to be preserved and the BTC felt that was enough to represent Gresley's work in the National Collection. Pegler was certainly not alone in feeling that an A3 ought to be saved, but he was the only person who could go to his bank and get the cash to pay for it. "There was never any question of buying any other engine."

Alan Pegler bought *Flying Scotsman* to save it from being scrapped, not because he had a desire to own a main line steam engine. But he had not bought it just to save it; he had bought it to keep it running. Had he expected to make money out of it? "My goodness, no!" Alan not only paid £3,000 to buy *Flying Scotsman*, there were then ongoing payments to BR to maintain and operate it, although income from railtours was obviously set against this. Pegler was quite unequivocal. The engine did not make money, "and anyone who thinks they can make money out of owning a main line steam engine is living in cloud-cuckoo land".

Having become the proud owner of a Gresley Pacific, Pegler needed to keep it somewhere, and he used his contacts within BR management to his advantage. At the time, Dr Beeching was the BR chairman, and his reorganisation of the railways went beyond pruning the system. BR workshops were being hived off to separate companies. On Alan's behalf, Beeching's number two, Sir Stuart Mitchell, asked the new works manager at Doncaster if he could accommodate a privately owned steam engine. "The poor chap was just started in a new job, and the BR vice-chairman came on the phone. I don't suppose he could really have said no." So, Alan got a small shed in the corner of Doncaster works where he could keep his engine.

Clearly the high point of *Scotsman*'s first period of main line running in preservation was the King's Cross – Edinburgh nonstop anniversary run in 1968. By then, No. 4472 had been running with a second tender for a couple of years, but the extra tender had not been purchased with this run in mind. The thinking was that it would make such runs as Leeds to Carlisle over the Settle & Carlisle route possible without a water stop.

Even with the two tenders, the Edinburgh nonstop was a close-run thing. Was there any opposition within BR management to such an ambitious undertaking? "No, I simply paid BR up front and, if anyone tried to stop it happening, I never heard of it," said Alan. "In fact, bets were being placed within BR management as to whether we would reach Edinburgh nonstop or not – but I don't think any of them would have nobbled the water troughs.

"If we thought we could not make Edinburgh, we were to whistle as we approached Berwick … unfortunately, a chap was lying on the platform trying to photograph the train, and the driver whistled at him. This whistle was misinterpreted as meaning we wanted to stop for the emergency water at Berwick, so we were signalled into the loop. Fortunately, we were able to keep going."

The return was much easier: "We found a chap with a key to the water supply at Scrooby, and got him to hold up the ball-cock so the troughs were filled to overflowing. After a good water pick-up, we ran 140 miles to King's Cross and arrived, nonstop from Edinburgh, with 2,000 gallons to spare."

It was the publicity surrounding the nonstop run that led to the American trip. It was an old family solicitor who actually suggested it. Alan had contacts in America, though, including Graham Claytor, president of the Southern Railway, but formerly vice-president (law). The law was to be very influential on this trip. Nelson Blount, a disciple of Billy Graham, agreed to sponsor it, but died in a flying accident immediately prior to *Scotsman*'s departure from the UK.

With hindsight, Alan admitted he probably should have cancelled the trip as a result but so many arrangements had been made, everything was in place and everybody was looking forward to it. Being vacuum-braked, the engine had to take its own train. There was a law that restricted 'foreign' locomotives and trains to circus or exhibition trains only on US railroads, so passenger carrying was out of the question. Alan also took a couple of Pullman cars, which were never part of his train, but he had agreed to deliver them to a museum in Wisconsin, and what better way could there have been?

Once over there, the US authorities were soon saying to Alan: "It's your engine; why don't you drive the goddamn thing?" So, while in Texas, he took a test for the benefit of the appropriate departments of New York State and Ottawa and, subject to supervision by a qualified traction inspector, Alan Pegler could finally drive his own engine. Les Richards, formerly BR North Eastern Region traction inspector, was part of the crew accompanying the train and, under his supervision, Pegler reckoned he drove No. 4472 over approximately 17,000 miles of US and Canadian railroads.

It was the law that finally put a stop to the adventure. "I knew I was going to go bankrupt, but I was enjoying myself," said Alan, but when it was pointed out that he was actually risking arrest, he took the threat very seriously. He was effectively trading while knowingly insolvent and aware that he was running up considerable bills that might never be paid — this is illegal both in the US and in Britain. Alan knew it had to stop.

He returned to Britain and filed for his own bankruptcy while George Hinchcliffe did what he could to secure the engine's future. William McAlpine had always been a keen supporter of the engine, and acted so quickly that No. 4472 was on a boat heading for the Panama Canal before any of the US creditors knew what was happening. Alan Pegler lost everything, but he worked his passage across the Atlantic as a ship's entertainer, a career he took to very quickly, as P&O even asked him to do it again. He was proud he was able to discharge himself from bankruptcy relatively quickly and "with creditors receiving a reasonable proportion of their money". He was also proud to have obtained his Equity card at the age of 60.

Was he pleased that *Flying Scotsman* is now at the National Railway Museum at York, preserved as part of the National Collection? "Yes, undoubtedly. It is where it always should have been." In fact, he was happy to be quoted as saying that, if he had cancelled the American tour and retained ownership of the engine, he would have donated it to the museum by then. He felt the engine's future as a working engine is much more secure in the museum's hands.

OVER THE POND

What could follow a London-to-Edinburgh nonstop steam run, 40 years on from the first, but now on an all-diesel route? There was one more challenge though: just as other celebrities, such as The Beatles and Oasis, once they become stars at home, then felt obliged to conquer America, Gresley's A3 Pacific No. 4472 set out to do just the same.

Flying Scotsman emerged from an overhaul at the Hunslet Engine Company in Leeds but after a few more runs it went into Doncaster works to be fitted with a bell, a large chime whistle and a cowcatcher. The cowcatcher was out-of-gauge for BR, so it was temporarily removed and, on August 31, 1969, No. 4472 had one last run from King's Cross to Newcastle before setting off on its biggest adventure yet.

The plan was for the locomotive to tour the United States and Canada, hauling a nine-coach exhibition train promoting British industry. To say that this plan was ambitious is something of an understatement. American railroads were never nationalised, so No. 4472 would have to run on several companies' tracks. Steam was eliminated much earlier than in Britain, and main line steam preservation was virtually non-existent in the 1960s.

There were few qualified steam crews and no steam infrastructure. Steam firemen with any knowledge of coal-fired engines would be even thinner on the ground. Even passenger trains were few and far between, and the engine would have to run colossal mileages on lines that only ever witnessed the passage of multi-engined container trains.

Flying Scotsman left Liverpool Docks on September 19 aboard the MV *Saxonia* for the ten-day journey to Boston, Massachusetts. Once the tour was under way, diesel and electric piloting was necessary some of the time but nevertheless No. 4472 immediately started to hit the headlines. The crew that accompanied the engine had to do everything, and that included driving it, something even Alan Pegler had not achieved (officially) on BR.

The initial tour, starting in Boston on October 8, 1969 and heading south down the eastern seaboard via New York and Washington to Atlanta, proved a great success, and No. 4472 travelled on to Slaton, Texas for the winter. The Southern Railroad was particularly pro-steam and coordinated the tour programme with the assistance of four other railroads. The engine proved reliable, and there were no reports of it delaying freight trains on long, single-track lines.

A similar tour was then planned for 1970, covering middle America and crossing the border into Canada on August 20, 1970. This was less successful as the British government was now actively discouraging companies from supporting the train, considering that the steam engine was giving a bad impression of British industry. The BR anti-steam attitude of the time had clearly spread even higher. But *Flying Scotsman* pressed on and completed the second stage of the tour, its owner thoroughly enjoying himself, even though he knew by then it could only end in bankruptcy.

After almost a year out of action, this time spent in the roundhouse at Spadina shed, Toronto, there was a glimmer of hope that the financial position could be salvaged by a period on show in San Francisco, starting with a trade fair known as 'British Week'. It was a 4,500-mile journey from Toronto to San Francisco, on routes that had not seen a steam engine since the early 1950s. In more recent years, main line steam has made a spectacular comeback on some very long-distance excursions in North America, and effectively it was *Flying Scotsman* that paved the way for these, by proving that it was still possible.

If only the train could have carried passengers, there might have been no financial problems, but it was found to be illegal to do so. In September 1971, No. 4472 set off from Toronto for San Francisco. Crossing the Rockies in Canada, it ran short of coal and was assisted by five diesels at one point. It crossed the Columbia River at Wishram, spending some time on tracks once owned by the Great Northern Railroad. On through the Deschutes Canyon and Bend, Oregon, *Scotsman* then steamed down the spectacular Feather River Canyon route of the Western Pacific Railroad, once used by the 'California Zephyr'.

On March 18, No. 4472 started a season of weekend passenger trips on the San Francisco Belt Railroad, the first steam and the first passengers the line had seen since the Second World War. After a promising start, standing in a prime position at Fisherman's Wharf, the train was forced to move to a much less satisfactory site and income dried up. Bankruptcy was now inevitable, and Alan Pegler even had to borrow the money for the air fare back to London to file the petition himself.

Perhaps his greatest achievement under the circumstances was that he managed to return to San Francisco and arrange for the safe keeping of *Flying Scotsman*, at a US Army base at Stockton, near Sacramento. Alan's only chance of returning to Britain then was to work his passage on a ship, and he started a new career as an entertainer on board a P&O cruise ship, a career that lasted seven years and enabled him to discharge himself from bankruptcy.

Flying Scotsman's rise to fame first time round was virtually assured once it acquired its name, as was its further fame once Alan Pegler had reincarnated it 40 years later, but the engine's story is littered with 'might-have-beens' ever since, and it is interesting to consider what might have happened if Pegler had not negotiated an agreement to run his engine on BR, taking the form of a legally binding written contract. Pegler's close contact with senior BR management put him in a unique position to achieve this, but BR was careful not to enter into any such agreement with anyone else. Without the agreement, *Flying Scotsman* would have become subject to the October 1967 steam ban and, after a couple of years' enforced idleness, would probably have gone to America anyway, as it had been under consideration for several years.

More interestingly, though, is what might have happened if, having the benefit of his unique running agreement on BR, Pegler had not been given the opportunity for the American tour, or if he had turned it down. No. 4472 would no doubt have continued running on BR right up to the end of the contract in 1971. Coincidentally, it was in that year that GWR 4-6-0 No. 6000 *King George V* made its ground-breaking tour of the Western Region, which led to the lifting of BR's steam ban.

Completely new rules were enforced for steam running from 1972, rules which *Flying Scotsman* was subject to on its return from overseas.

The major difference was that No. 4472 had simply continued running in much the same way after purchase in 1963 as it had previously, with BR taking full responsibility for its operation on what was still, at least partially, a steam-worked railway system. As steam was phased out in different parts of the country, and steam infrastructure removed, BR found the operation of *Flying Scotsman*, and the other privately owned steam engines that briefly saw main line use in the 1960s, progressively more difficult.

In reality, by 1969, even No. 4472 was quite restricted as to where it could actually run, only its second tender making it a viable proposition at all. Had *Flying Scotsman* still been operating in Britain in 1971, *King George V*'s experimental runs would have been unnecessary, and the engine might not even have been returned to steam by Bulmers. In fact, *Flying Scotsman* might have been adopted by Bulmers instead. With hindsight, BR's main line steam ban was sensible, and No 4472's temporary emigration was timely. No one really knows whether BR intended the ban to be permanent; no doubt some board members did and some did not.

One thing is certain, *Flying Scotsman* would never have become the icon it is now if the ban had remained in force, and we must all be grateful to the many people who worked so hard to get the ban lifted as soon as it was practicable.

Chapter 11

WILLIAM McALPINE

✳✳✳

ILLIAM MCALPINE, a director of the well-known civil engineering firm Sir Robert McAlpine Ltd, with its many railway
connections, came to the rescue. The situation called for a
man with enthusiasm, money and the right contacts, and Mr McAlpine
was just the man for the job.

It was not a case of buying *Flying Scotsman* from Pegler: he was
bankrupt, so the engine belonged to his creditors. McAlpine had to
pay off these creditors quickly and discreetly and get the engine moved
out of the US before either a known creditor took it upon himself to
sell No. 4472 for scrap, or another creditor surfaced and demanded even
more money. McAlpine achieved the seemingly impossible remarkably
quickly and, on January 19, 1973, *Flying Scotsman* was his, and was on the
MV *California Star* bound for Liverpool via the Panama Canal.

McAlpine was an enthusiast and an owner, not only of steam locomotives, but of an entire railway in his garden. Garden railways are not
particularly unusual, and some are even full-size narrow-gauge systems,
but McAlpine's, the Fawley Hill Light Railway, was standard gauge!

As from 1972, when steam returned to Britain's main lines, locomotive
owners took responsibility for the operation of their engines. It was

their duty to provide BR with an engine in full working order, coaled and watered ready to run, and to ensure that it was coaled, watered and serviced as necessary during the day.

All BR did was provide the crew, the train and the track to run on. Pegler had not had to do this initially; he had simply paid BR to run the engine, although a team of volunteers led by George Hinchcliffe had become increasingly involved in its operation. Now, in effect, BR paid the engine owners to hire their engines (only a nominal sum, of course).

This was a radically different approach for the post-steam era. Under McAlpine's ownership, *Scotsman* needed a team of willing volunteers to accompany the engine and attend to its every need during the day.

McAlpine was one of four millionaire private owners, or part-owners, of *Flying Scotsman*, but his ownership was the most enduring, yet at the same time the most low-key of them all. He never featured on the six o'clock news, posing on the footplate of 'his' engine. He simply owned it, and it continued to do its job, but at times spectacularly. In fact, McAlpine contributed far more to railway preservation than is generally realised, and continued to do so up to his death in 2018. He had eventually given up ownership of No. 4472 only because he felt that its future would be more secure with a younger part-owner at least.

Mr McAlpine became Sir William in 1980. Although his ownership of *Flying Scotsman* was quiet and low-key, it was eventful, and his 20-year involvement saw the engine's fame increase enormously.

✳ ✳ ✳

If William McAlpine had not secured *Flying Scotsman's* release from the US by paying off the engine's creditors, those creditors would have sold it eventually, to recover part of their debt. It is perhaps surprising that they did not do so immediately, although possibly this was because the engine was considered to be of little value in the US at the time. Scrap value in Britain might have been £10,000 and in the States probably a similar figure.

Although there was a market in railway memorabilia in America, it was nothing compared to the UK today, where a nameplate alone can fetch £60,000. The debts incurred were no doubt way in excess of

£10,000, so it would never have been worthwhile for any of the creditors to seize the asset and sell it, as they would have been forced to share the proceeds with everyone else; and, as is the case in most bankruptcies, they might have received the equivalent of .001p-in-the-£1 each. The danger is that it only needs one creditor to threaten to take action and demand a separate deal to undo all that has been achieved. It has only become known recently that the figure paid in settlement of the debts plus transport was £25,000.

The engine might have been sold for scrap, or to a private collector, or museum, or to an American preserved railroad. However, the attraction of a British steam engine in America is not all it might be expected to be. There are two LNER A4 Pacifics preserved the other side of the Atlantic, and neither museum has ever tried to steam its engine in the expectation of picking up rich profits from American tourists.

It was perhaps the tour promoters' over-estimation of the fame of the engine and its money-making potential that caused the venture to fail spectacularly, but it was also the reason why it was possible to save the engine relatively painlessly.

If it had passed into American ownership, like the SS *Queen Mary*, it would have returned to the UK by now. Any private collector would have died eventually, with his collection dispersed. A railroad operating it for profit would have found it uneconomic by comparison with US-built engines and would have offered it for sale. *Pendennis Castle* went to Australia, and SR Schools class 4-4-0 No. 30926 *Repton* went to Canada, both as operating engines, but both were sold and eventually repatriated. The same would have been true of *Flying Scotsman*, provided it had not been scrapped. The story may not have been vastly different if William McAlpine had not stepped in, but that is not to understate the debt that British steam enthusiasts owe to him.

SCOTSMAN COMES HOME

When *Flying Scotsman* returned to the UK in February 1973, it was to a very different world from the one it had left behind in 1969. The BR ban on main line steam had come into force in October 1967, but No. 4472 had been exempted, because of Pegler's binding contract with BR. Now though, that had expired.

Fortunately, the steam ban had been short-lived, but *Flying Scotsman* was now to be subject to exactly the same rules and regulations that applied to all the other, lesser engines. It would be allowed on just six routes, in spring and autumn only, and there would be no exceptions to the rules.

However, No. 4472 was still regarded as something of a special case and, if rules could be bent to ensure the engine kept its dignity, they would be. After a year of disuse and weeks standing on the deck of a cargo vessel in Atlantic storms, it was inspected at Liverpool and passed fit to run. *Flying Scotsman* made its way light engine to Derby under its own steam, hardly bothering with technicalities such as 'approved steam routes', and entered the one-time LMS locomotive works for an overhaul. Emerging as good as new in July 1973, it was time for it to start earning its keep as quickly as possible.

It had emerged in slightly more authentic condition than in Pegler's day. The nameplates were black, the cylinders were black (though now with red lining) and the second tender was blue and grey to match BR coaching stock, although it was now rarely used. The number 4472 had reappeared on the cabsides, and *Flying Scotsman* ran virtually unchanged in visual appearance for 20 years.

Steam remained banned from BR tracks in summer, but again No. 4472 at least travelled under its own power, hauling two vintage saloons also acquired by McAlpine, the destination being Paignton, for a busy summer season on the Torbay Steam Railway. It was only when this was successfully completed and main line action could be entertained again in September that the A3 could get back into the railtour business, under the new regulations.

For maximum impact, No. 4472 teamed up with the engine that had finally managed to break BR's ban, GWR 4-6-0 No. 6000 *King George V*, an engine that, coincidentally, had travelled to America in 1927 and carried a commemorative bell. The train was billed as 'The Atlantic Venturers Express' and ran from Newport to Shrewsbury on September 22, 1973.

No. 4472 was then back into promotional duties again. No. 6000 was based at Bulmers' cider factory in Hereford, along with an exhibition train formed of Pullman cars. The exhibition train was to tour the country promoting Bulmers' products and, although steam haulage was favoured, *King George V* was 'out of gauge' for the vast majority of the BR system, so *Flying Scotsman* stood in, although still restricted to the 'approved routes', such as York to Scarborough and Newcastle to Carlisle.

No. 4472 needed a new home, as Doncaster diesel depot was no longer available. The Hon John Gretton, of Stapleford Park, near Melton Mowbray, who had become the owner of *Flying Scotsman*'s one-time arch-rival, GWR 4-6-0 No. 4079 *Pendennis Castle*, was setting up a railway centre at Market Overton in Rutland, and this was chosen as the engine's home base, at least for the winter months.

Market Overton was at the end of a disused branch line from the East Coast Main Line at Highdyke, north of Stoke Tunnel, which had served ironstone mines. There was a small but fairly modern locomotive shed that had housed the quarry's fleet of industrial diesels. Other engines followed, including the National Railway Museum's ex-Barry scrapyard SR Merchant Navy Pacific No. 35029 *Ellerman Lines*, which was to be cosmetically restored but sectioned to show the internal workings of a steam engine, in the new museum to be opened at York. Considerable engineering skills were being established at Market Overton, which would bear fruit in the future.

As a preserved railway, though, the scheme never took off. It was an inconvenient location for main line operations, remote from all the 'approved routes', and would never be suitable in itself as a venue for big engines to stretch their legs. *Flying Scotsman* and *Pendennis Castle* quickly deserted their new home and made for Carnforth in the north-west.

LOADING GAUGE

A loading gauge was designed to ensure that people loading open wagons did not stack the goods so high or wide that they would strike over-bridges or tunnels. Every goods yard had one, and many can still be seen today. The term is also taken to mean the dimensions of a locomotive or item of rolling-stock. All must be designed to fit comfortably under all bridges and tunnels they are expected to encounter.

The loading gauge on Britain's railway system varies more widely than might be expected, but it is certainly among the smallest in the world. The fact that British steam locomotives hold so many world records is all the more surprising, taking into account that they had to be built so much smaller than in every other country.

Broadly speaking, ex-Great Western main lines have the largest loading gauge, especially in terms of width, as they were originally built to 7ft 0¼in gauge. The Great Central London extension also has a generous loading gauge, having been built relatively late and designed to accommodate Continental-sized rolling-stock at a future date. Scottish lines tend to be narrow, and some of the tightest clearances are in Kent.

Flying Scotsman and the other early A1s immediately fell foul of loading gauge problems. They were designed to haul Great Northern Railway expresses between King's Cross and York but, once the GNR was absorbed into the LNER, it would be much more useful if they could run on all LNER main lines. They had to be trimmed slightly to fit under bridges on the North Eastern Railway, and slightly more for the North British routes north of the border.

The LMS main lines have a slightly more generous loading gauge and, once cleared for all LNER main lines, the A1s could run on nearly all main lines in Britain. This has proved particularly useful in more recent years as *Flying Scotsman* has been able to run in most parts of the country. An exception remains Kent, where the only express engines that can be used on most routes are those designed by the Southern Railway.

GWR engines are not generally much taller and wider than their counterparts, although 4-6-0 No. 6024 *King Edward I* has been shortened in recent years to give it wider route availability away from GWR main lines. The problem with GWR engines is the wide front bufferbeam and the distance between it and the leading driving wheels. The overhang on a sharp curve through a station platform has caused engines to become stuck when away from home territory, even in recent years. Gresley's A1s, although more compact and with much less front-end overhang on curves, would still have struck platforms on sharply curved Scottish routes. This led to corners being cut out of the bufferbeams — just one difference between Gresley's original design and *Flying Scotsman* today.

Unfortunately loading gauge is not constant. Diesels have generally been designed to be 'go anywhere' machines and there have been well-publicised instances where track has been reballasted, reducing clearances to an extent that remains perfectly acceptable for diesels but becomes too tight for many steam engines, and at least three engines have struck bridges in recent years.

In fact, BR, Railtrack and Network Rail must be commended for (usually) ensuring that clearances are not reduced, just in case a steam engine ever wants to use the route. There have been occasions when this has not been possible, though, particularly on electrified routes, where there must understandably always be a minimum clearance between the tops of chimneys and wires carrying 25,000 volts.

Axle loading is also a problem, of which the A1s quickly fell foul when built. It is not the total weight of the engine that is the issue, it is the weight carried by each axle, which can vary widely according to where the weight of the engine is concentrated. Gresley was given a maximum figure to work to, but in practice the A1s still exceeded the weight that had been calculated, despite using as many weight-saving ploys as possible. Axle loadings have generally been increased, and *Flying Scotsman* can now run on many routes that would have been unthinkable in days gone by, particularly private heritage lines on quite lightly-engineered branch lines.

Chapter 12

APPROVED
STEAM ROUTES

* * *

G WR 4-6-0 No. 6000 *King George V* toured part of the Western
Region in October 1971 with the Bulmers' cider train, plus
additional coaches for fare-paying passengers. As a result,
BR announced the end of its infamous steam ban, with effect from
June 1972.

Operations though, were to be strictly limited to just six short routes,
in spring and autumn only, and for just a few engines based on or near
the designated routes.

Gradually, the BR system was opened up to steam, culminating in
September 1992 with the whole of the Southern Region, only at night,
but then in daylight from the following year. Busy inter-city routes,
especially 25kv overhead electrified territory, remained strictly no-go,
and routes were deleted as electrification was extended.

Flying Scotsman joined in from the second half of the 1973 season,
and quickly stretched its legs on a fair proportion of the designated
routes, often with the first train to use a particular route, for example
Liverpool – Manchester in March 1980.

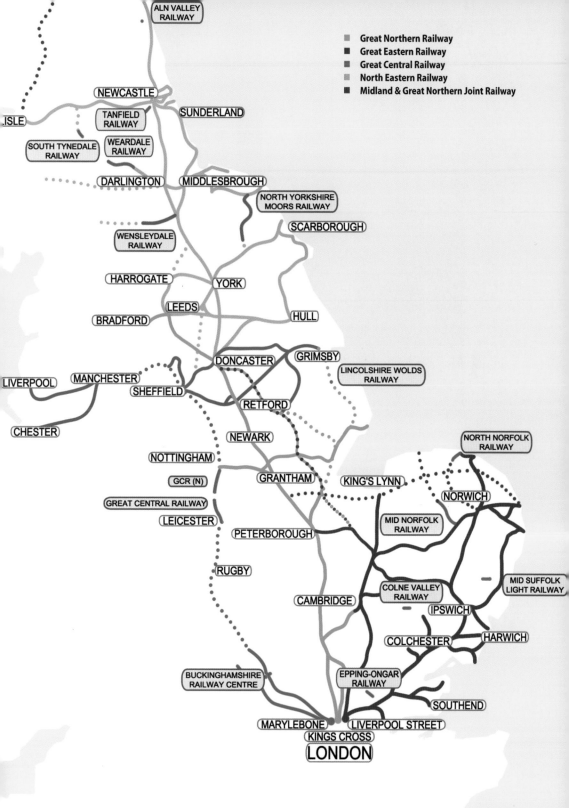

The Great Northern Railway's main line ran from King's Cross to Doncaster, with routes to Cambridge, Nottingham, Skegness and Grimsby. It also served Leeds and Bradford. The map shows the routes remaining open in 2016, with the Grimsby line and other routes in Lincolnshire having closed by 1970.

GNR Stirling Single 4-2-2 No. 1, in steam at Loughborough on the Great Central Railway in December 1981.

Herbert Nigel Gresley.

A typical Gresley-designed teak-bodied coach on the North Yorkshire Moors Railway.

Flying Scotsman, Gresley's best-known creation.

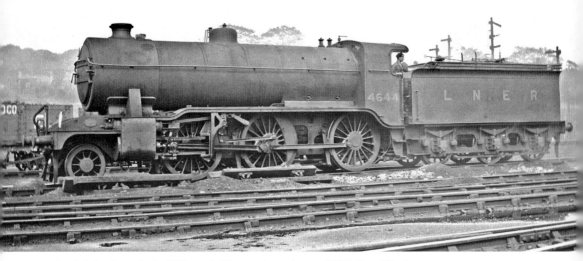

Gresley's first design for the GNR was the H1 2-6-0, later to become LNER K1 and K2 class. PAUL CHANCELLOR COLLECTION

Gresley's GNR 2-8-0 heavy freight engine. PAUL CHANCELLOR COLLECTION

Gresley's third A1 Pacific No. 1472 at Doncaster shed soon after construction in 1923. The engine, still unnamed, was allocated here initially and visits to King's Cross were comparatively rare at first as the Pacifics, with their eight-wheeled tenders, were too long for the turntable until it was extended. W.H. WHITWORTH, RAIL ARCHIVE STEPHENSON

Patrick Stirling's 4-2-2 No. 1. Although the 'single-wheeler' was used by many of Britain's pre-Grouping railways, it was Patrick Stirling of the GNR who took the concept to its ultimate form. These were remarkably powerful engines for their size and could haul trains of 90 to 100 tons from Grantham to King's Cross in two hours. In theory, speed was limited to 60mph on the Great Northern, in common with most railways at the time, but one of Stirling's engines, a 2-2-2, is known to have reached 86mph. Withdrawn and preserved by the Great Northern Railway in 1907, No. 1 was steamed again by the LNER in 1938, and by the National Railway Museum in 1981. It is seen on the Great Central Railway at Loughborough in May 1982.

H.A. Ivatt's pioneer Atlantic 4-4-2 No. 990 *Henry Oakley* at Boston in 1998. No. 990 was the first engine of this wheel arrangement to run in Britain and it reversed most of Stirling's design principles. The small cylinders actually result in No. 990 being theoretically less powerful than Stirling's No. 1, but the extra wheels made the engine much better at starting heavy trains and the much bigger boiler meant it was able to maintain its power for much longer. Externally, though, many of the GNR's design principles were continued, with No. 990 still having the same flowing lines in the footplate as well as an overall tidy and sleek appearance. Ivatt's large Atlantics followed, an equally graceful but much bigger design.

A3 4-6-2 No. 4472 *Flying Scotsman* at Carnforth. Although changed in appearance in more recent years, the Pacific design when new in 1922 still continued the GNR tradition of compact and graceful design. Gresley's engines proved to be some of the fastest ever, but they were also powerful for their size when compared with later and bigger British Pacifics. Despite its size, *Flying Scotsman* still has the flowing footplate inherited from Stirling's 8ft Single, as well as many other design features, often considered to be among the best in British steam locomotive design practice. JEFF COLLEDGE

Stephenson's *Rocket* was an 0-2-2-wheel arrangement, with the bare minimum number of wheels, which gave it speed rather than power.

The predecessors of Gresley's Pacifics were the Atlantics, referring to a 4-4-2-wheel arrangement. GNR No. 251 is one of H.A. Ivatt's later designs, with a much larger boiler than the earlier Atlantics.

No. 4472 *Flying Scotsman* is a Pacific; a 4-6-2. The front four-wheel bogie under the cylinders guides the engine smoothly round curves, and helps support the heaviest part of the engine. Six coupled driving wheels can transmit more power to the track than four wheels. Although eight driving wheels would give more power still, this was never successfully used in Britain for express locomotives as the engines were just too long and rigid for the curves on British main lines, although eight driving wheels became standard for express engines in many other countries such as the US. The rear two wheels under the cab were not an articulated pony truck on a Gresley Pacific, but a Cartazzi truck, where the frames were rigid right to the rear of the engine, with side-play in the axle boxes to allow for sideways movement on curves.

The LNER and LMSR came into existence on January 1, 1923. The LNER's *Flying Scotsman* stands alongside the LMSR's Stanier Pacific No. 6201 *Princess Elizabeth*, the first Pacific design introduced by the competing LMS, but not until as late as 1933.

The GNR standard goods engine, the J22 0-6-0 became the J6 in LNER classification. J was retained by the LNER from the GNR as denoting 0-6-0. The numbering for the classification normally ran through GNR, GCR, GER, NER, NBR and GNSR in that order, with LNER-built classes taking later numbers, e.g. J72 and J94. There were exceptions to this general rule though. PAUL CHANCELLOR COLLECTION

Flying Scotsman heads the 'Flying Scotsman'. No. 4472 passes Palmers Green in 1929 on its way to make the *Flying Scotsman* publicity film. RAIL ARCHIVE STEPHENSON

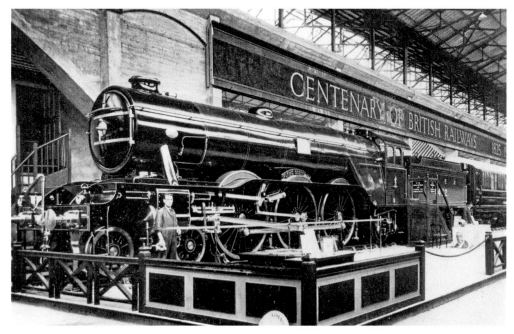

A1 Pacific No. 4472 *Flying Scotsman* on display at the British Empire Exhibition at Wembley in 1925 with a standard LNER six-wheeled tender.

BR introduced 'The Caledonian' in 1957, running between Euston and Glasgow. Initially hauled by Stanier's Princess Coronation Pacifics, the impressive headboard was in use into the diesel era.

The locomotive used on the Southern Railway's 'Golden Arrow' boat train was the most highly decorated ever to run in daily service in Britain, with a full smokebox headboard, British and French flags on the bufferbeam and even a huge golden arrow on the side of the locomotive.

Three varieties of LNER Pacific on display at Barrow Hill Roundhouse. On the left is the 2008-built Peppercorn A1 No. 60163 *Tornado*. No. 60532 *Blue Peter* was one of a later batch of Peppercorn A2 Pacifics, introduced in 1947. The Peppercorn A1s and A2s were externally similar apart from their driving wheel diameters. Gone are the external design touches, such as the smooth-flowing footplate and wheel splashers from the Gresley era. The whole design is more functional and angular. Smoke deflectors were fitted to all new Peppercorn A1s and A2s. On the right is Gresley A4 Pacific No. 60009 *Union of South Africa* in BR green livery.

In BR Brunswick green livery, as carried for 13 years but as now turned out after its latest overhaul, No. 60103 *Flying Scotsman* prepares to depart from King's Cross on January 14, 1963 on its last run for British Railways, the 1.15pm to Leeds. The engine was purchased for preservation by Alan Pegler who is seen standing on the bufferbeam. NATIONAL RAILWAY MUSEUM

Flying Scotsman attracted large crowds wherever it went. During the 1967–72 BR steam ban, Tyseley in Birmingham was one of the few places where main line steam engines could operate, but only over a few hundred yards of track. Only *Flying Scotsman* was allowed actually to haul trains on the main line, and had arrived at the head of a railtour. TERRY ROBINSON

The crew of *Flying Scotsman* at Boston, Massachusetts, about to steam across America.
TERRY ROBINSON

The unmistakeable figure of Alan Pegler at the Railfest celebrations at York in 2004. Alan died on March 18, 2012 at the age of 91. ROBIN JONES

Flying Scotsman in the US at night. TERRY ROBINSON

Almost a British scene at Boston at the start of the tour as No. 4472 comes face to face with a London bus. TERRY ROBINSON

William McAlpine did not just buy *Flying Scotsman*: he also collected and restored a number of historic coaches at Carnforth. This train was often used for entertaining corporate clients, sometimes with No. 4472 at the head. Here it passes Bolton, then little changed since steam days, on September 28, 1979, with a McAlpine private charter from Carnforth to Dinting, near Glossop.

One of the six original 'approved' routes for steam was the Hope Valley line between Sheffield and Manchester. On a clear autumn afternoon on November 10, 1984, No. 4472 *Flying Scotsman* rounds the curve at Buxworth. The train was 'The Fenman', which No. 4472 would haul as far as Spalding, the route beyond Sheffield having been given one-off dispensation, as often seemed to happen with this particular engine, on its way to the East End for a Royal assignment.

Carnforth shed retained its steam age atmosphere well into the preservation era. On March 21, 1976, *Flying Scotsman* stands in the yard with Barclay 0-4-0 crane tank Glenfield, while Dr Peter Beet's German Pacific No. 012-104 steams past.

No. 4472 departs from Sellafield station adjacent to the nuclear power station, returning to Carnforth on May 30, 1987.

No. 4472 *Flying Scotsman* backs onto the 'Scarborough Spa Express' at York on July 29, 1981.

From 1972 to 1978, BR would not allow steam trains to run in summer, because it could not spare any coaches at weekends. Steam was not permitted in winter for reasons that are less clear; from 1979, this restriction was lifted and wintertime steam has been a great success, leading to the return of such spectacular sights as *Flying Scotsman* topping a snowbound Ais Gill summit on the Settle & Carlisle line on January 30, 1983.

Flying Scotsman on the turntable at Marylebone on September 28, 1986.

Heading for Australia on a trip involving a round-the-world journey, *Flying Scotsman* is lifted onto the deck of the appropriately named SS *New Zealand Pacific* on September 11, 1988. GEOFF SILCOCK

Shortly after sunrise on October 20, 1988, No. 4472 passes Culcairn in New South Wales, just south of Wagga Wagga, where the annual Eisteddfod was being held. The engine is en route from Sydney to Melbourne for the Aus Steam 88 steam festival.

Flying Scotsman at Alice Springs in August 1989. NICK PIGOTT

In BR Brunswick green, with double chimney and smoke deflectors, No. 60103 *Flying Scotsman* passes Waterside on the Paignton & Dartmouth Railway on August 12, 1993.

Its LNER apple green livery having been applied at the eleventh hour, the reborn No. 4472 *Flying Scotsman* – complete with double chimney – stands proudly at King's Cross on the morning of Sunday, July 4, 1999 while passengers board the historic 'Inaugural Scotsman' train to York to mark its comeback. ROBIN JONES

Roland Kennington pauses briefly from his labours on *Flying Scotsman* in Southall shed. ROBIN JONES

Sir Richard Branson celebrates the saving of *Scotsman* with one of several bottles of bubbly. NATIONAL RAILWAY MUSEUM/PA

Pupils of Inglebrook School hand over their pocket money to the National Railway Museum's head of education, Julia Fielding, in the museum's Great Hall.

Flying Scotsman in the works at the National Railway Museum, partly dismantled, but on public display in 2005.

The A3 boiler carried by *Flying Scotsman* from 1978 to 1995, which was to be rebuilt by Riley & Son Engineering and has now been replaced on the engine during the museum's first overhaul.

Ian Riley, head of Riley & Son Engineering, of Bury. JACK BOSKETT

Heritage Painting's Mike O'Connor and his daughter Teriann, 21, apply finishing touches to the cab after the application of the 60103 numbers on February 17, 2016. ROBIN JONES

No. 60103 *Flying Scotsman* is back on the Great Northern main line on February 25, 2016 heading the inaugural train from King's Cross to York past Holme, after completion of its ten-year restoration.

Former *Flying Scotsman* driver Ron Kennedy with his daughter Katie, on board the inaugural run. Ron, who now lives in Leigh-on-Sea, said: "I never dreamed about being on it again. To be out with it is just fantastic. It belongs to the public really so let's keep it so it's always there for our children and grandchildren." TIM ANDERSON/ NATIONAL RAILWAY MUSEUM

Network Rail released this picture on February 25, 2016 issuing a warning to spectators to say off the tracks. NETWORK RAIL

Flying Scotsman and its crew are mobbed by waiting crowds at York. ROBIN JONES

LNER A3 Pacific No. 60103 *Flying Scotsman* departs from Goathland on the North Yorkshire Moors Railway on March 16, 2016.

Flying Scotsman crosses the Royal Border Bridge at Berwick-upon-Tweed on May 14, 2016 with Steam Dreams' 'Cathedrals Express' from London to Scotland.

No. 60103 *Flying Scotsman* climbs past Borthwick on the newly opened Borders Railway on May 15, 2016. This train was nearly unable to run steam-hauled over the line as Network Rail had not carried out the necessary gauging requirements for the locomotive.

No. 60103 and No. 45407 depart from Rawtenstall on the East Lancashire Railway with a public train on January 10, 2016.

LNER A3 Pacific No. 60103 *Flying Scotsman* crosses Batty Moss viaduct at Ribblehead on March 31, 2017 with a railtour from Oxenhope on the Keighley & Worth Valley Railway, to Carlisle marking the reopening of the Settle & Carlisle line as a through route.

No. 60103 arrives at Oakworth on the Keighley & Worth Valley Railway on April 8, 2017.

Flying Scotsman runs alongside a High-Speed Train and a Class 91 electric on the four-track section of the East Coast Main Line at Beningborough, north of York, on April 23, 2017 during the making of a film promoting the introduction of new 'Azuma' units on ECML services.

LNER A3 Pacific No. 60103 *Flying Scotsman* departs from Lincoln southbound with the Railway Touring Company's 'Scarborough Flyer' to King's Cross on June 24, 2017.

Flying Scotsman tops Ais Gill summit on May 22, 2018 with the Edinburgh–Carlisle–York leg of Steam Dreams' 'Cathedrals Express' tour from London to Scotland.

No. 60103 departs from Lincoln northbound with the King's Cross-to-York leg of the Railway Touring Company's 'Great Britain XI' tour on April 19, 2018.

LNER A3 Pacific No. 60103 *Flying Scotsman* passes LNER A1 Pacific No. 60163 *Tornado* at Wansford on the Nene Valley Railway on September 27, 2018.

Flying Scotsman approaches Newark on January 12, 2019 with 'The Scotsman's Salute', a tour from King's Cross to York which ran in memory of Sir William McAlpine, one-time owner of *Flying Scotsman*.

Waiting to greet the train on platform 9 at York was a human flying Scotsman in tartan-clad bagpiper Pipe Major David Waterton-Anderson from Carlton, near Selby. He said: "I'm also a private pilot so I'm literally a Flying Scotsman. I have ridden on this train as a youngster." ROBIN JONES

It was 1987 before the A3 made it to the Southern Region though; it rarely ran on the Western Region and never beyond Bristol, and perhaps most surprisingly its only visits to Scotland in 30 years of operation were Mossend – Perth and Carlisle – Ayr runs in October and November 1983, the latter diesel-piloted between Carlisle and Annan.

In 1972, the original routes were Didcot – Tyseley, Newport – Shrewsbury, Carnforth – Barrow-in-Furness, York – Scarborough, Newcastle – Carlisle and Guide Bridge – Sheffield. In 1973, this was expanded to include Barrow – Sellafield, Hull–Scarborough, Leeds – Carnforth and Inverkeithing – Dundee. 1974 saw the following added: Basingstoke – Westbury and Edinburgh – Perth – Aberdeen. 1975 one-offs for the Rail 150 celebrations at Shildon were Battersby – Whitby and Sheffield – York – Newcastle. In 1976, routes were considerably expanded to include Shrewsbury – Chester, Newcastle – Middlesbrough replaced Newcastle – Carlisle, Leeds – York via Harrogate, Perth – Inverness, Inverness – Kyle of Lochalsh, Sheffield – York and Manningtree – March. 1980 saw in Liverpool – Manchester, Manchester – Leeds via Standedge and Manchester – Blackburn – Hellifield. 1984 saw in Fort William – Mallaig and Marylebone – Banbury. In 1985, added as one-off were Bristol – Plymouth, Gloucester – Swindon and Plymouth – Truro. 1986 saw in Salisbury – Yeovil. 1987 saw in Glasgow – Fort William, Chinley – Buxton and Machynlleth – Aberystwyth/Pwllheli.

By 1994 steam was permitted on a wide variety of routes but with the exception of the main inter-city ones, particularly those that were electrified. Privatisation was to see a completely new situation though.

STEAMTOWN CARNFORTH

The unremarkable little Lancashire town of Carnforth occupies a unique position in the hearts of British steam enthusiasts, as it was one of the last three BR steam locomotive sheds to operate steam, right up to the last day of scheduled services on August 3, 1968.

By then, a collection of preserved engines was already stored in the yard, with a view to their future use on a steam railway being proposed in the Lake District. The renting of the whole shed and yard after the end of steam was planned to be a temporary arrangement, until the line could be secured and reopened. Further engines arrived, including several LMS 'Black Five' 4-6-0s purchased privately at the end of steam.

The line from Plumpton Junction on the Cumbrian Coast line, through Greenodd and Haverthwaite to Windermere Lakeside would have been an excellent proposition for a steam railway, but the revivalists were unable to prevent half the line being swallowed up by the A595 Haverthwaite bypass, severing the remaining short section from the BR system. Many of the Lakeside & Haverthwaite Railway Society members pressed on with the project, and the three-and-a-half-mile Lakeside & Haverthwaite Railway has developed into a very professional and attractive heritage line. Carnforth shed, marketed as Steamtown, also developed into a visitor attraction in its own right, and many of those involved, feeling that the Lakeside branch could never compare with what was originally planned, stayed put and continued to develop Steamtown.

The lifting of BR's steam ban in 1972, and Carnforth's nomination as a main line steam operating base, sealed the shed's future. It then started to attract more and bigger engines looking for main line action and, as early as 1973, this included *Flying Scotsman*, which adopted Carnforth as its permanent operating base from October 1974. It was an association that was to last for almost 15 years.

Today, Carnforth shed serves a different, but no less valuable purpose. The railway restoration and engineering side of the business was expanded, and eventually purchased by the West Coast Railway Company, set up by Yorkshire farmer, businessman and steam locomotive owner David Smith.

The engineering business had already dealt with contracts such as the refurbishment of the Pullman coaches for the Venice Simplon Orient-Express. After the privatisation of BR, WCRC became a Train Operating Company in its own right, operating for example, the 'Jacobite' summer steam service between Fort William and Mallaig. The train operating side of the business has also steadily expanded and WCRC operated the 'Ride the Legend' York-to-Scarborough trains hauled by *Flying Scotsman* during the summers of 2004/5.

The shed is not now open to the public, though. Its facilities never made it an ideal venue for visitors and its upgrading could never justify the cost. Carnforth though, still plays a vital part in keeping steam alive on Britain's main lines.

CUMBRIAN COAST EXPRESS

One of the first routes to be sanctioned by BR for occasional steam operation from 1972 was from Carnforth to Barrow-in-Furness, giving the small fleet of LMS 'Black Five' 4-6-0s preserved at Carnforth an opportunity to stretch their legs occasionally.

In 1973, the approval was extended farther up the Cumbrian coast to Sellafield and when *Flying Scotsman* first visited Steamtown at Carnforth in September 1973, it made its first-ever run on this route – the first of what was to be a very large number of runs.

In late 1974, No. 4472 returned to Carnforth, this time on a permanent basis, and by now it was permitted to run not only to Sellafield, but south to Leeds. The West Coast Main Line over Shap, and the nearby Settle & Carlisle line remained out of bounds, but the Cumbrian Coast line to Sellafield was a very attractive alternative.

The 64-mile route was a reasonable distance and could be covered each way with steam in a day, even with a train originating from Euston, as most railtours did. It is scenic, with the sea on one side and the Cumbrian Fells on the other side all the way. Although it is relatively level, there is a short, sharp climb out of Ulverston one way and Lindal on the return to test the engine.

While Sellafield is not an ideal destination for a day trip, there is also Ravenglass a few miles farther south, where most passengers would alight for a trip on the Ravenglass & Eskdale miniature railway. The engines could be turned at Sellafield on a triangle at the British Nuclear Fuels plant, or alternatively on the Ministry of Defence Vickers gun range system south of Ravenglass. In practice, engines did not take kindly to the curves on either of these triangles and it was usual to use two different engines, with one running light engine tender-first between Carnforth and Sellafield.

Flying Scotsman's appearances on the route grew more frequent and such was its popularity that, on one occasion in May 1976, a railtour comprised no less than 18 coaches. The engine was honoured with the assistance of LNWR 2-4-0 No. 790 *Hardwicke* as far as Ulverston, where more substantial assistance was provided in the shape of LNER B1 4-6-0 No. 1306 *Mayflower*.

In July 1978, though, the unthinkable happened and BR itself started to run regular steam trains for the first time in ten years. It hired *Flying Scotsman*, which had just emerged from an overhaul at Vickers Ltd in Barrow, where it was fitted with its spare boiler, the one Alan Pegler had purchased in 1963. It was, in fact, an A4-type boiler once carried by No. 60019 *Bittern*.

The train was the 'Cumbrian Coast Express', starting from Blackpool and aimed at holidaymakers. It was steam hauled from Carnforth to Sellafield and back, using No. 4472 one way and LNER A4 Pacific No. 4498 *Sir Nigel Gresley* the other. It was to run every Tuesday throughout the summer, but was very quickly expanded to Wednesdays as well.

The programme was extended in 1979 and other engines were included, but various circumstances led to what had been a very successful venture being surprisingly short-lived. It did, however, prove that regular timetabled main line steam trains could make money.

Steam operation continued in the form of occasional one-off tours, and was extended again, as far as Maryport, with engines able to continue light to Carlisle. The trains could go no further as there are limited clearances on the Maryport-to-Carlisle section and it was feared that passengers leaning out of windows could suffer serious injury.

The regular service was relaunched in a slightly different form in 1987, but this time the train was called the 'Sellafield Sightseer', the destination quite unashamedly being the nuclear power station, which had opened a visitor centre and wanted to improve its public perception. Again *Flying Scotsman* was part of BNFs marketing 'spin'.

The trains were successful, but controversial, attracting not quite the type of media interest that had been envisaged. Unfortunately, it virtually spelt the end of steam working on the Cumbrian Coast line for some years, owing to the increased publicity surrounding the perceived health hazards of the area.

Carnforth and the Cumbrian Coast were a long way from *Flying Scotsman*'s normal sphere of operation in its regular service days, but it proved that the engine was nothing if not adaptable, and this was one of the engine's most successful periods in its preservation career.

THIS IS YORK

Flying Scotsman has a long association with the city that is now its permanent home. Gresley designed his A1 Pacifics to run between King's Cross and York and for all of their working lives, the A1s and later the A3s ran mainly on the East Coast Main Line passing through the city.

During its days fitted with a corridor tender, working the Newcastle and Edinburgh nonstops, *Flying Scotsman* would have been an almost daily sight in York, but once the A4s took over the top duties in 1935, No. 4472, as a King's Cross engine, would rarely have worked north of Grantham. Only in its very last days, with the benefit of the double chimney and Kylchap, were the Grantham engine changes largely abandoned, and King's Cross A3s used on runs right through to York and Newcastle again.

Once in preservation, up until 1969 the East Coast Main Line remained a favourite route for *Flying Scotsman*, based not far away at Doncaster. On its return from the US though, No. 4472 found itself with only a handful of routes to run on and, although York – Scarborough was one of them, the engine used the route only once, in 1972, hauling the Bulmers Cider exhibition train.

Based at Carnforth from 1974, *Scotsman* was allowed to run to Leeds but no further. 1975 was a special year though. The Stockton & Darlington centenary celebrations took place at Shildon at the end of August, followed by the opening of the new National Railway Museum at York. This led first to a procession of steam engines passing through York en route to and from Shildon but, perhaps better still, a short series of railtours on the East Coast Main Line.

Flying Scotsman made the journey from Carnforth to Shildon on an August Sunday afternoon, hauling just one coach, William McAlpine's Caledonian Railway observation car No. 41, and piloted by none other than the LNWR's 1892-vintage 2-4-0 No. 790 *Hardwicke*.

After its appearance at Shildon, when it hauled NER 2-4-0 No. 910 in the famous cavalcade on August 31, No. 4472 returned to Carnforth with a lengthy empty stock train, this time accompanied by GWR 4-6-0 No. 6960 *Raveningham Hall*.

Two weeks later *Scotsman* returned many of the coaches to York, this time piloted from Carnforth by the National Railway Museum's LNER Gresley V2 2-6-2 No. 4771 *Green Arrow*. Nowhere in Britain had seen such a level of steam activity for many years, and it was a strong indication that the new museum at York was about real working steam; it was not going to be just a static collection of inanimate objects.

The four main line railtours in September 1975 were planned to feature four different LNER Pacifics, *Flying Scotsman* naturally being one of them. In the event, A4 No. 4498 *Sir Nigel Gresley* worked the first one, from Newcastle to York and back, while the remaining three were shared by the apple-green trio of B1 4-6-0 No. 1306 *Mayflower*, V2 *Green Arrow* and No. 4472 in various combinations.

Then the following year, it was back to normal, although Leeds-to-York-via-Harrogate was added to the list of 'approved' routes. *Flying Scotsman* immediately showed that it was still a bit of a special case, by using the main line between Leeds and York instead on April 24. In fact, running via Harrogate was so inconvenient that the direct route was soon authorised for regular steam, as was the York-to-Sheffield line, giving No. 4472 much easier access to York and other parts of the BR system.

In 1979, after the success of the 'Cumbrian Coast Express' from Carnforth, BR introduced its own 'York Circular' steam service, initially running twice daily on a York, Harrogate, Leeds, York itinerary. Steam runs to Scarborough were very occasional, particularly after BR lifted the turning triangle at Filey, but in 1981 the ex-Gateshead shed turntable was installed at Scarborough, paid for by Scarborough Borough Council, and BR itself began running the 'Scarborough Spa Express', a 212-mile York-Harrogate-Leeds-York-Scarborough and return operation, three days a week. Both York- and Carnforth-based engines were used, including *Flying Scotsman*.

SETTLE & CARLISLE

On May 1, 1976 LNWR 2-4-0 No. 790 *Hardwicke* again teamed up with *Flying Scotsman* for a run with a train of historic coaches to commemorate the centenary of opening of England's most ruggedly spectacular main line and a firm favourite with steam enthusiasts. It was an absolutely foul day and the train was not allowed to run steam-hauled on the Settle & Carlisle line itself.

The steam section was from Carnforth to Hellifield, where the engines were removed but were at least permitted to travel as far as Settle station light engine, while the train ran diesel-hauled to Carlisle.

This unspectacular event was the first small step towards the return of steam to the route that had been the talk of the enthusiasts' world for years. The problem was that, from the six routes approved for steam in 1972, the list had grown steadily, but inter-city main lines and electrified routes were off limits. While the West Coast Main Line over Shap would never have been approved anyway, it was electrified in 1974 and this effectively put Carlisle station out of bounds, so although the Settle & Carlisle could well have been given approval in 1974 or 1975, the overhead wires now seemed to preclude it forever.

In every other country in the world, steam has operated on electrified routes without a problem and, as electrification was extended in Britain in the 1960s, this had resulted in more and more instances when a steam engine ran into an electrified station. Eventually it was agreed by BR that it was still safe to do so in the 1970s, under certain conditions. This was a crucial decision as without it, steam operation would be extremely limited in Britain today.

At Easter 1978, steam returned to the Settle & Carlisle line, the engine leading the way being *Flying Scotsman*'s smaller sister, V2 2-6-2 No. 4771 *Green Arrow*. No. 4472 itself made it up the 1-in-100 southbound to the 1,169ft Ais Gill summit on June 16 that year, with a McAlpine private charter train.

It was not until September that the engine hauled a public passenger train on the route. On this occasion two trains ran, using three engines: the A3 together with SR Merchant Navy 4-6-2 No. 35028 *Clan Line* and BR 9F 2-10-0 No. 92220 *Evening Star*. The trains ran in memory of perhaps Britain's most famous steam photographer, Bishop Eric Treacy, who had collapsed and died at Appleby while photographing *Evening Star* in May that year.

After a handful of steam trips in 1978, the Settle & Carlisle was ruled out for steam traction during 1979 after the collapse of Penmanshiel Tunnel, which resulted in East Coast Main Line expresses using the route. From 1980 though, steam returned to the line on an almost weekly basis with the inauguration of the 'Cumbrian Mountain

Express', a train that frequently featured haulage by *Flying Scotsman*.

Once the S&C was opened up for steam, it became the favourite route for enthusiasts, engine owners, railtour operators and the public. The threat of closure in the 1980s enhanced its popularity and since its reprieve in 1989, passenger and freight traffic on the line has expanded rapidly, to the extent that it has sometimes become quite difficult to fit a relatively slow-moving steam train in between the other traffic.

Nevertheless, the West Coast Railway Company in recent years has operated a busy programme of steam trains over the line on behalf of a number of railtour promoters, particularly in summer. *Flying Scotsman* returned to the line on February 6, 2016 on its first public main line railtour since its latest restoration. A severe landslip that day resulted in the line closing as a through route for 13 months but *Flying Scotsman* was chosen to head a railtour on March 31, 2017 marking the reopening of the line. It has also hauled several trips from York to Carlisle in the last three summers for the Railway Touring Company plus occasional Steam Dreams' 'Cathedrals Express' trains over the line.

RETURN TO 'THE SMOKE'

Flying Scotsman was built for the King's Cross to York main line, and spent most of its main line career based at King's Cross shed. Built at Doncaster and preserved by Alan Pegler at Doncaster; in later years it was regarded as a Yorkshire engine and was regularly seen at York, but it had always been very much a London engine in 'real' steam days.

Early 1983 saw the engine's 60th birthday, and considerable rule bending took place to allow *Flying Scotsman* to celebrate this milestone in an appropriate manner by running on the Great Northern main line. One trip, with the engine hauling the train one way from Grantham to York, was envisaged initially, but demand was such that one run quickly became three and better still, the engine changing point was altered to Peterborough, allowing the A3 to be reacquainted with Stoke Bank for the first time in 14 years.

The anniversary of emerging from Doncaster works was celebrated by a private run from Carnforth to York, but the engine then ran with a couple of coaches to Peterborough late on a Saturday evening, to turn on the loop installed for flyash trains in the 1960s.

On three consecutive Sundays from February 27, No. 4472 took over its well-filled Pullman trains at Peterborough and headed north, the third trip actually running to Carnforth as opposed to York.

The trains were popular with the travelling public but much too popular with lineside observers. While other steam engines attract a good turnout of enthusiasts, not all of whom strictly obey the rules of trespass, but who are generally sensible, *Flying Scotsman* attracts everyone and his dog. Its fame can sometimes be its undoing. Trespass by the public reached almost epidemic proportions and East Coast Main Line expresses were stopped in order to clear people from the tracks.

All three trains ran without loss of life or limb, on the railway at least, but any thoughts that BR management might have had about permitting steam on inter-city main lines on a regular basis in the future quickly evaporated.

There was never any question at the time of *Flying Scotsman* or any other engine running in or out of King's Cross and the behaviour of the public in 1983 suggested that this was a very wise policy decision on the part of BR. Visits to London by No. 4472 had continued up to its American trip in 1969, but after that it was almost unknown in the capital for many years. It did appear at an exhibition at Kensington Olympia on August 10/11, 1974, and clocked up another notable feat by taking the exhibition train overnight to Carnforth afterwards, on routes where steam was then supposed to be forbidden.

However, No. 4472 did have a quick visit to the East End in November 1984, for a Royal Train no less. On November 20, for the official opening of North Woolwich station museum, No. 4472 hauled a short Royal Train from Stratford, conveying HM the Queen Mother, who performed the opening ceremony. *Flying Scotsman* added to its fame by becoming the first preserved steam engine to haul a Royal Train on the main line.

Regular steam passenger trains at any London terminus simply did not happen for more than 15 years but steam's sphere of operation continued to expand and eventually in 1985, official sanction was given for steam to use Marylebone station. The one-time terminus of the Great Central Railway's London extension was a quiet backwater compared with other London termini, but it had been a regular haunt of *Flying*

Scotsman during the war and in the early 1950s, and in 1985 it could even still boast an operating turntable.

First to break the new ground was Flying Scotsman's Carnforth stablemate, A4 Pacific No. 4498 *Sir Nigel Gresley*, the great designer's 100th Pacific named in his honour in a ceremony at Marylebone station in November 1937. While at first it was enthusiasts' tours, often steam-hauled one way, the plan was for a regular series of steam-hauled upmar-ket dining trains to Stratford-upon-Avon, using engines drawn from a small fleet stationed in the diesel multiple unit depot by the station.

Flying Scotsman moved to Marylebone in December 1985, and took its turn on 'Shakespeare Limited' trains to Stratford during the year, but in 1987 the engine found itself moving around the country to an unprecedented extent, still nominally based at Carnforth and working in the north of England, but also sometimes hauling trains from London, or in the Midlands, or even on the Southern Region between Salisbury and Yeovil, a route also recently opened up for steam.

At the end of 1987, though, *Flying Scotsman* moved south on a more permanent basis, not to Marylebone, but to be based at Southall in west London. This was to spell the end of its long association with Carnforth. Roland Kennington took over as the engine's chief engineer, and he immediately had a big job to do. After being absent from the capital for many years, *Flying Scotsman*'s return to working trains out of London was apparently only going to be for a very brief period. It may have found a new home in west London, but that was not going to be its operational area, at least not for the next 12 months or so.

Chapter 13

DOWN UNDER

***

S EVERAL BRITISH steam engines have been to North America, and
a few have immigrated to Australia, but only one has ever visited
America *and* Australia. 1988 saw the bicentenary of the first arrival
of British ships in Australia and in true Aussie tradition, some major celebrations were called for. Quite apart from all the other parties, Aus Steam
88 was arguably the biggest steam spectacular ever staged anywhere in the
world and in recognition of the country's origins, it was felt that a famous
British steam locomotive would add a certain something to the occasion.

The Australians officially asked for *Mallard* from the NRM, but it was
bad timing in that, although it happened to be in steamable condition,
it was the 50th anniversary of its world speed record in 1988, and it was
felt that *Mallard* ought really to stay in Britain. However, a Melbourne
postman took the story forward.

Walter Stuchberry was the chairman of the Aus Steam 88 committee
and he was determined to get a famous British steam engine to the
event. He asked Sir William McAlpine for *Flying Scotsman* instead and
it was agreed, subject to financial guarantees, that No. 4472 could attend.

Aus Steam 88 was very much a live steam event, centred on Melbourne,
and engines travelled to it under their own steam, not just for exhibition

but to haul trains in an intensive two-week programme of railtours. Australia consists of six mainland states, and these are far more independent than might be expected, even to the extent of having largely separate railway systems built to different gauges. New South Wales in the east adopted standard gauge. Victoria in the south and South Australia chose broad (5ft 3in) gauge, while Queensland and Northern Territories followed African tradition, with 3ft 6in gauge. In fact, one-third of South Australian railways were built to 3ft 6in gauge.

In recent years, there has been a programme of standardisation, with standard gauge track being extended from New South Wales, parallel with the broad gauge across Victoria to Melbourne, and parallel with 3ft 6in track to Brisbane in Queensland. The lines to Alice Springs, and across the Nullarbor Plain to Perth, have been converted to standard gauge, a process completed in 1969. A bonus of having broad- and standard-gauge lines running parallel for hundreds of miles is that it is possible to run two steam trains alongside each other on parallel tracks, as the Australians frequently do.

Although *Flying Scotsman* was initially scheduled simply to sail to Melbourne, go on exhibition for two weeks, haul a couple of railtours, then stay on for a few months afterwards and work more tours from Melbourne and Sydney, there was potential for something rather more ambitious, potential that simply had not existed a few years earlier, before the spread of the standard gauge network.

Roland Kennington had overhauled the engine at Southall remarkably quickly, including the fitting of air-brake equipment; it had a run to Stratford-upon-Avon on one of the dining trains and was pronounced fit to emigrate, sailing from Tilbury on September 11, 1988 on the P&O container ship *New Zealand Pacific* via the Cape of Good Hope. While en route, a slight problem arose as the Port of Melbourne had sold the only floating crane able to lift an A3 Pacific off the deck of a container ship, and *Flying Scotsman* had to be offloaded at Sydney instead on October 16, just days before the start of the big party 300 miles away in Melbourne. No. 4472 had to be steamed virtually the minute it touched Australian soil and, after a quick test run, set off on a two-day journey to Melbourne.

The Aus Steam 88 organisers had a cunning plan. Although *Flying Scotsman* was to be one of many steam engines on display in Melbourne, it was expected to be such a popular attraction that it was totally enclosed in a cocoon within Spencer Street station, and the public had to pay to go inside to see the Gresley Pacific. It worked, as 130,000 people paid to see it and receipts during the two weeks virtually covered the engine's transport costs.

After Aus Steam 88, *Flying Scotsman* stayed in Australia to haul tours farther afield, back in New South Wales, working from Sydney, and then double-heading a 12-day Sydney-to-Brisbane tour using the new standard gauge line. Its partner was Australia's most famous Pacific, No. 3801, a semi-streamlined engine that has itself visited most parts of the continent. *Scotsman* was so popular and performing so well that it was agreed to extend its stay, and plans were formulated for even more ambitious outings.

In August 1989, *Flying Scotsman* set out from Melbourne for the run to Alice Springs, the first steam engine to use the new standard gauge line. It was almost impossible not to break records, and No. 4472 covered the desolate 422-mile stretch of outback from Parkes to Broken Hill, nonstop in nine and a half hours, a world record for a nonstop run by a steam engine, beating its own record from 1928, which it had repeated in 1968.

For the Australian run, two extra water 'gins' were coupled behind the tender, standard practice for Australian steam trains in a land with virtually no water. There were no half-full water troughs, fire hydrants or standby road tankers on the line to Alice Springs, and the engine simply had to haul sufficient water for the journey, as well as a heavy trainload of passengers and a diesel 'just in case'.

Flying Scotsman was not the only British steam engine in Australia, as there was already another British express steam engine in operational condition. In 1925 it had stood next to *Flying Scotsman* at the Wembley Exhibition and claimed to be the most powerful steam engine in the world. In comparative tests it had proved superior in many ways to *Flying Scotsman,* and many years later it had shared a shed with No. 4472 in deepest Rutland. Now though, *Flying Scotsman* was the star, and no one had thought it worthwhile moving *Pendennis Castle* across Australia for Aus Steam 88.

GWR 4-6-0 No. 4079 had been sold to the Hamersley Iron Railway back in 1977 and was occasionally used for special trains on what is

93

otherwise a totally heavy freight railway. The thought of reuniting the two British engines obviously occurred to a few people. If *Flying Scotsman* could get to all the other state capitals, it could get to Perth. It would be a gigantic undertaking, even to move *Pendennis Castle* 1,000 miles to Perth by road, but it was possible, and once a sponsor was found for the road transport, the reunion was on. *Flying Scotsman* crossed the Nullarbor Plain, with its 299 miles of dead-straight track, becoming the first steam engine ever to cross the whole of Australia under its own steam, and on September 17, 1989, it came face to face with No. 4079 at Perth. The two engines embarked on a programme of tours in the area, sometimes double-heading, and sometimes hauling separate trains on parallel tracks, something that can be done in some areas, even on lines of the same gauge.

This was a fitting climax to a hugely successful tour of no fewer than 28,000 miles of virtually trouble-free running. If *Flying Scotsman* was not already the most famous steam locomotive in the world, it certainly was now. The engine returned to the UK on the French vessel *Le Peruse*, from Sydney, this time via Cape Horn, becoming the first steam engine to circumnavigate the globe.

When *Flying Scotsman* returned from Australia, arriving on December 14, 1989, it was obviously tired from its exertions, but it still had a valid BR main line running certificate. It was soon back in action, and simply resumed a similar programme to the one it had left three years earlier, in marked contrast to its return from America almost 20 years earlier.

The main line certificate had only two years to run though, and not only was consideration being given to the next major overhaul, but privatisation of British Railways was looming, and the future for main line steam operation looked more than a little uncertain.

In a slight departure from its normal sphere of operation, No. 4472 even paid a working visit to a preserved steam railway, the Severn Valley, in September 1990. Prior to this, the engine's only serious runs on a preserved line had been on the Torbay Steam Railway in Devon in 1973, when it had returned from America and been unable to run on the main line immediately, simply because it was summertime. It is hard to believe that Gresley's A1s, when built, were almost too heavy even for their own main line, yet in 1990, *Flying Scotsman* could run on a privately owned ex-GWR cross-country secondary route.

Preserved steam operations have always centred on a boiler certificate for insurance purposes, which lasts for ten years, after which a total strip-down, inspection and overhaul is compulsory. BR, though, imposed a shorter, seven-year limit on its main line certificate and, by the 1980s, with weight restrictions being lifted, owners of main line engines were finding that, once their main line certificate had expired, they could run the engine for another three years on private heritage lines, to raise money for the next overhaul.

By 1992, this was even looking like a sensible option for *Flying Scotsman*. Many steam lines wanted it, believing it could be a big money-spinner, and for No. 4472's owner it would see them through the approaching nightmare of privatisation, after which a decision could be taken on the engine's main line future. The seven-year main line certificate expired after a run to Stratford-upon-Avon on October 25, 1992.

Chapter 14

PRIVATISATION OF BRITISH RAILWAYS

***

No. 4472 retired from the main line on October 25, 1992 and almost immediately embarked on a tour of preserved lines, visiting the Great Central and East Lancashire railways as well as the Birmingham Railway Museum at Tyseley, where 'driver experience' courses were proving extremely popular — and what better than driving *Flying Scotsman*? Other railways were queueing up to hire the engine, among them the Llangollen Railway.

No. 4472 duly arrived at Llangollen and promptly failed with serious boiler problems. Back in 1963, Alan Pegler had returned the engine to its single-chimney form, in LNER livery as he remembered it. It was never really an issue at the time, as it was not called upon to run as fast or as economically as it had needed to do when in daily express train haulage. After all, the single chimneyed No. 4472, still in low-pressure form as an A1, had officially broken the 100mph barrier in 1934, so in single-chimney A3 form as Doncaster had returned it to in 1963, was still a good engine. It never reverted to being an A1, let alone a GNR-style A1 so, even in Pegler's day, it was a bit of a hybrid.

For those involved in running the engine though in later years, whose memories tended to be from the 1950s and 1960s, there was a desire to see *Flying Scotsman* as they remembered it, and in its optimum mechanical form, the fitting of the double chimney in 1958 having made a huge difference. Roland Kennington, in charge of No. 4472, had been quietly collecting the parts necessary for the transformation, in the knowledge that he would struggle ever to get Sir William McAlpine's agreement to it. The premature boiler failure at Llangollen meant that at least a boiler overhaul was necessary to keep the engine running, but the full main line overhaul was still not justified. This led to a surprise development in 1993, when *Flying Scotsman* appeared in its final BR condition as No. 60103, in Brunswick green, with double chimney and Kylchap, and even smoke deflectors.

It was the talk of the steam enthusiasts' world, even though most passengers were disappointed with many unable to believe that it was the real *Flying Scotsman* that was hauling their train. McAlpine had agreed to it only until it could return to the main line. A double-chimneyed A3 could haul heavier trains more economically, but the transformed engine was only going to run at 25mph, mostly on GWR branch lines, hauling six or seven coaches.

The transformation took place with the assistance of the engineering company, Babcock Robey, which carried out the boiler overhaul, and *Flying Scotsman* emerged from its works in July 1993 as No. 60103. It immediately resumed its intensive programme of preserved line running, returning first to the Paignton & Dartmouth (formerly Torbay) Steam Railway, now looking (and sounding) very different. This was followed by the Swanage, Gloucestershire Warwickshire, Llangollen, Nene Valley and Severn Valley railways.

It was while No. 60103 was on its tour of heritage lines that a significant change of ownership occurred. It was rumoured that McAlpine had sold half of *Flying Scotsman* to Pete Waterman, but this is not strictly true. McAlpine owned the operating company, known at various times as SLOA Railtours, Pullman-Rail and Flying Scotsman Services; its assets included Steamtown, the locomotive shed at Carnforth, as well as *Flying Scotsman* and assorted rolling stock. What was sold to Waterman was a share in the company.

In the run-up to privatisation, British Railways' train operations had been split into Inter-City, Regional Railways, Network SouthEast, Rail Express Systems and Railfreight. These became autonomous businesses, which were to be sold to the private sector.

One other, smaller business, was the Special Trains Unit, which was a subsidiary of Inter-City, and owned several sets of coaches used for special trains. Steam trains formed only a small part of this business, which also hired out its coaches for football specials, diesel railtours and any other special train working anywhere in Britain. When *Flying Scotsman* hauled a train, Scotsman Services often hired the coaches from the Special Trains Unit, which was responsible for running the train. Now the Special Trains Unit, headed by David Ward, was for sale to a private investor.

Well-known pop impresario and entrepreneur Pete Waterman bought it but, at the same time, McAlpine and Waterman agreed to merge their two companies. Now *Flying Scotsman* was the figurehead of all special train operations on the about-to-be-privatised railway system — except, of course, that it was not certified to run on the main line. Nevertheless, there had been considerable investment in ensuring that there would be main line steam under the privatised regime and, with *Flying Scotsman* now effectively owned jointly by two extremely successful millionaire entrepreneurs, the future looked good.

* * *

British Railways was privatised on April 1, 1994, but it was not a return to the pre-1923 situation of private companies owning trains and track. A new organisation, the late-lamented Railtrack, owned the track and infrastructure and new Train Operating Companies (TOCs) were set up to operate train services, buying 'paths' on the system from Railtrack.

Steam operations had previously come under the InterCity banner, through its Special Trains Unit, which dealt with all railtour and charter train operations, although it did not have a total monopoly and Network SouthEast and Regional Railways had occasionally run their own steam trains.

The most far-reaching effect of privatisation on the operation of main line steam was a result of the principle of 'open access'. A TOC had the right to operate a train on any route providing it was safe and practicable, regardless of motive power. Suddenly BR's policy of 'approved routes' for steam was out the window. Theoretically steam could now go anywhere – and quickly did.

The politics of privatisation grew very complicated, and *Flying Scotsman*, without turning a wheel on the main line, was well and truly embroiled in the politics.

Pete Waterman had become involved as there was potential for new organisations to make money out of the newly privatised railway system and Waterman saw an opportunity not to be missed. He had bought the InterCity Special Trains Unit and called it Waterman Railways and by owning all the rolling stock used on all railtour operations, this would give him a lucrative business, especially if he also owned some of the engines that would pull the trains. He had already achieved a minor miracle by getting a diesel locomotive passed by BR for main line operation at colossal expense, but no other diesel ever earned this accolade, and privately owned diesels did not get a fair crack of the whip until privatisation.

Waterman Railways was never a TOC and its trains were operated by Res (Rail Express Systems) whose charging policies were causing problems for all tour operators. Waterman Railways went through some organisational changes with Flying Scotsman Railways becoming a new parent company, while Waterman Railways was the rolling-stock hire company, but access charges led to the company withdrawing from promoting its own trains. It would only hire trains to other railtour promoters.

Flying Scotsman failed again with boiler troubles at Llangollen in 1995 and returned to Southall. With its ten-year certificate expiring shortly, it was now time for a decision as to whether to embark on the full main line overhaul. In view of the bad experience of the first months of Waterman Railways' operations, it was clear that the plans for *Flying Scotsman* to be the centrepiece of Waterman Railways' railtour operations, were not going to happen, as Waterman Railways no longer had any railtour operations of its own.

If Pete Waterman's plans had materialised, *Flying Scotsman* would probably have been overhauled to main line standards, possibly retaining its BR green livery and double chimney. It would have been a great success with many of Britain's steam enthusiasts, but this would never have paid the bills, as the overhaul would have cost a large seven-figure sum. Waterman is above all a businessman, and if the engine was not going to make money, it would not be part of his strategy. If, and it is a big if, it had been overhauled by him once to main line condition, it might have run for seven years, but would more probably have been sold. There is now a market for main line steam engines at the right price, and anything could have happened to the engine if Flying Scotsman Railways had sold it as a going concern.

Chaper 15

THE ENGINEER'S TALE

✳✳✳

GEOFF COURTNEY talked to Roland Kennington, Scotsman's chief engineer from 1986 to 2004. There are several names synonymous with *Flying Scotsman*, particularly in its preserved, post-BR life. Roland Kennington is one of them. Roland is the man who for nearly 20 years lovingly kept No. 4472 going against what were, at times, overwhelming odds. Not for him, or his dedicated team of volunteers, warm and well-equipped premises, with an endless flow of money and a plethora of spare parts. Instead he spent nearly two decades with the locomotive in dark, dank and cold sheds – and they don't get darker, danker or colder than Southall motive power depot on a January night – with clapped-out machinery, empty coffers, and a demanding public expecting the LNER legend to look and perform like a thoroughbred racehorse. And he loved every emotionally draining minute of it.

Roland was chief engineer for *Scotsman* from 1986, when Sir William McAlpine was the owner, until its sale bv Flying Scotsman plc to the National Railway Museum in 2004. His involvement resulted from one of the many mini crises that have punctuated the locomotive's preservation history.

Just two days before Christmas in 1985 he received a telephone call at the Bedford engineering company W.H. Allen, for whom he was production manager. On the line was Ray Towell, the man in charge of No. 4472 at Carnforth, where it was then based. Ray had a problem that revolved around a broken combination lever, the Christmas holiday, and Sir William's 50th birthday. "Basically the lever had snapped in two, on a test run up north, following an overhaul," said Roland. "It turned out the problem had been caused by a bolt being carelessly left in the motion during the overhaul. "I had been working as a volunteer on A4 No. 4498 *Sir Nigel Gresley* at Carnforth for the previous two years, so Ray thought of me when the problem arose. "I told him the works was closing the following day for ten days over Christmas, and he said something had to be done because the engine was needed for a special train early in January to mark Bill McAlpine's 50th birthday. The best I could offer was to weld the broken lever, which wasn't very satisfactory, but Ray eagerly agreed."

So, the two parts of the lever were brought down from Carnforth by train to Milton Keynes and then taxi to Bedford – surely the cabbie's most unusual passenger over that festive period – and Roland welded the parts together. "BR agreed to allow No. 4472 to run with the welded repair for two journeys only, but that was enough," said Roland. "Bill got his birthday train."

Four months later Roland took a call that would literally change his life. "It was Bernard Staite, who worked for Bill and was in effect in charge of *Flying Scotsman*. "He told me that Bill wanted to move the engine away from Carnforth and from the team that was then looking after it – I don't think the combination lever incident had done them any good – and place it in the care of a group of volunteers.

"Bernard said they were looking for an honorary chief engineer to head the group – note the word honorary – and my name had been mentioned. I was flabbergasted, and said there were two things to consider: firstly, I didn't know anything about steam engines, and secondly I needed to talk to my wife, Chris."

The first consideration was no problem to Bernard or Bill, and the second was no impediment either. "Chris told me I'd do what I wanted whatever she said, and anyway she didn't mind. I decided I could fit it in with my full-time job at Allen's, so the next day I called Bernard and agreed to do it."

Roland said the Friday before he was due to start his honorary, voluntary position he went on a run with No. 4472 and after its return to Marylebone, where it was then based, he overheard some traction inspectors talking about the testing of three-cylinder engines. "I thought to myself: 'You're out of your depth here, Roland.'" But Roland wasn't out of his depth. "Without sounding big-headed I discovered I knew more than I thought, thanks largely to my 'apprenticeship' with *Sir Nigel Gresley.*"

Having settled down into his role, events soon took another seismic shift. "Just before Christmas 1987 Bernard Staite told me that the state of Victoria wanted to borrow the locomotive as part of its celebrations to commemorate the Australian bicentenary, and that if it happened, he wanted me to accompany the engine. "To be honest, I didn't think it would come off, but it did. It was initially for six months, and to my surprise Allen's agreed to release me, albeit without pay. However, Bill McAlpine knew the company's boss, and Bill got him to agree to continue paying my salary while I was in Australia."

That six months away from home turned out to be 13, with Roland flying to Australia in September 1988 and meeting No. 4472 at Sydney where it arrived after six weeks on the high seas. The highlight of No. 4472's stay Down Under was undoubtedly the nonstop run on August 8, 1989 from Parkes to Broken Hill, a distance of 422 miles. The journey was accomplished in just less than nine and a half hours, with extra water being provided by three 7,000-gallon water carriers – or water gins in Oz parlance – behind the corridor tender. The run was the longest-ever nonstop journey by a steam locomotive, a world record which stands to this day and almost certainly will never be beaten.

Back home towards the end of 1989, the scenario for both Roland and No. 4472 took on a darker hue. "I was made redundant – Allen's probably thought that if they could without me for 13 months they could do without me for good – and money for the engine started running out."

Roland's outlook brightened up when he found work with another engineering company, but for *Flying Scotsman* money became tight.

By 1993 pop impresario Pete Waterman was involved. He and Sir William each owned 50 per cent of the company, which in turn owned the locomotive, with Pete the chief executive. "Bill let Pete run the company and he did what he liked," said Roland. "In 1993 the ticket for the main line ran out and private railways were invited to make bids for being loaned the engine. The result was a two-year programme on nine lines.

"This was a double-edged sword for me. I was out of a job by then, but was at least travelling around the country looking after the engine and getting my expenses paid, but at the same time everyone knew *Scotsman* was being beaten into the ground."

That beating took its toll, and the locomotive was eventually declared a non-runner on the Llangollen Railway, where it spent an ignominious spell out of steam with visitors 'cabbing' it at £1 a time. Roland was to discover, however, that there is indeed a silver lining behind every cloud. "A story about the engine's condition was broadcast on the radio, and as a result the Midlands boilermaking firm Babcocks offered to restore the boiler, in return for publicity. Bill gratefully accepted, and as I had no job at the time, he asked me to work with Babcocks.

"A new smokebox was made from BR drawings, and I grabbed the opportunity to fit a double chimney, which I had wanted to do for a long time. Babcocks also made some smoke deflectors from original drawings. After ten weeks the work was done and the private railway programme continued, but the Bill McAlpine ownership period was nearing its end.

"The company that owned the engine was seriously in debt, and Bill took back control and in 1995 put the locomotive up for sale. This was picked up by one of the railway magazines, and soon after I got a call from Bill saying he had sold the engine — and me with it!"

It transpired that pharmaceutical entrepreneur, Dr Tony Marchington, had seen the article and agreed to meet Bill's £1.25-million price without even inspecting the engine, which by then was in a sorry state in Southall shed, its base for several years. "I had never heard of Tony Marchington, so I rang him from a callbox on Waterloo station to introduce myself," said Roland.

"I arranged to meet him at Southall, where he would see for the first time something for which he had paid more than £1 million!" Tony immediately sanctioned a rebuild of the sorry-looking *Scotsman*, a job that was to take three years and cost not far short of another £1 million.

"I was elated," said Roland. "It meant I was going to be able to do something I had wanted for years – turn a mediocre locomotive into the equivalent of an A4. Tony wanted a 'Rolls-Royce' job, and I was determined to give it to him". And a Rolls-Royce job it certainly was. On July 4, 1999 *Flying Scotsman* made a triumphant return to the main line, when an estimated million people lined the route from London to York. The engine looked absolutely gorgeous in apple green and carrying the number 4472, and there were more than a few tears as youngsters and adults alike cheered and waved the mighty locomotive and feted its crew.

"For the first two years of Tony's ownership everything went well," recalled Roland. "The engine was very busy, and there was great interest everywhere we went. But then money became tight and I used every trick to keep it running, although it never ran in an unsafe condition. "We had to be ingenious at times."

Ingenuity wasn't enough, however, and after well-documented travails No. 4472 was once again put up for sale. In sharp contrast to that under Sir William McAlpine, this sale was carried out in the full glare of the public and media, with the National Railway Museum mounting an overt and unabashed publicity campaign aimed at securing the locomotive 'for the nation' while private individuals were also beavering away in the background.

Eventually the NRM won the day, paying £2.1 million plus buyer's premium. "There was a certain amount of inevitability about the NRM getting the engine," said Roland. "I was disappointed. I readily accept that all good things come to an end, but I do not think it should have gone to such an organisation or have been bought with Lottery money – I think it should have stayed in private hands. "That's not sour grapes, but I guess that is how it may sound."

No Roland, it doesn't. After 18 years of dedication to *Flying Scotsman* going way beyond the call of duty, you have a right to say how you feel. Every anorak, every enthusiast, and every member of the public who will in future enjoy this LNER masterpiece, is in your debt.

Roland Kennington passed away in 2017, aged eighty.

Chapter 16

FLYING SCOTSMAN PLC

***

*F*lying Scotsman's tour of heritage lines came to an abrupt end on April 23, 1995, when it was withdrawn from service with boiler troubles, again while at Llangollen. The 'dream team' of McAlpine and Waterman could have been good for the engine under different circumstances, but both seemed to agree that it needed more money spending on it than either was prepared to invest. Its earning potential simply did not appear to be sufficient to justify the expenditure required.

For the first time since 1972 in America, *Scotsman* was in limbo, its future uncertain. It needed another owner and that person came along in the larger-than-life figure of Dr Tony Marchington, proprietor of Oxford Molecular Group plc, a company with a value estimated at £200 million. In February 1996, Dr Marchington bought Flying Scotsman Railways, including the engine and all the company's other assets – which included a set of Pullman cars but not Waterman Railways – for £1.25 million, and embarked on probably the most thorough and certainly the most expensive overhaul carried out on a British steam engine up to that time.

Roland Kennington, who had been the engine's chief engineer since 1986 and who knew it inside-out, remained in charge, with the work being carried out in the one-time GWR locomotive shed at Southall. It was a job that was to take more than four years; the plan was that on completion, *Flying Scotsman* would undertake an intensive series of upmarket premium-priced dining trains for wealthy customers all over Britain.

Marchington invested part of his considerable private fortune in the project. A team of financial experts was assembled and the future looked rosy, although many steam enthusiasts recognised that *Flying Scotsman* was no longer part of their scene, as the engine's future trips would be aimed at a totally different price range, way beyond their means.

1995 had seen a very significant development when SR Bulleid Merchant Navy Pacific No. 35028 *Clan Line* had emerged from an over-haul fitted with air brakes, making it the first steam engine to be able to haul the premier luxury train, the Venice Simplon Orient-Express. This had proved to be a great success, and Marchington and his team felt that the market was big enough for another similar operation, but with the added attraction of Britain's most famous steam engine always at the head. Costs escalated, though, with first the Pullman cars being sold, so when *Flying Scotsman* did appear in the summer of 1999, it was going to have to haul coaching stock hired elsewhere, with a consequent effect on profits.

The overhaul was not just thorough; it amounted to a total rebuild into what was, in effect, a non-streamlined A4, with the boiler, which was from an A4 anyway, uprated to a pressure of 250psi, and the cylinders rebored. *Flying Scotsman* was more powerful than ever, still fitted with its double chimney and, in a major departure from previous practice, it was not only fitted with air brakes, but it had its vacuum braking system removed too.

It carried LNER apple green livery though as No. 4472, making it completely unacceptable as an authentically preserved locomotive. But this was not the idea; the engine had to earn its keep, and the instantly recognisable livery and number was part of the marketing plan, while the design enhancements would make it capable of hauling heavy trains anywhere.

Just as when No. 4472 disappeared to America and returned to a totally different world, completely different circumstances greeted No. 4472 on its return from the wilderness this time: the brave new world of privatisation. The main plank of privatisation was open access — Railtrack could not simply enforce arbitrary rules as to what could run where. If it was practicable it was permitted and since 1994, steam has been allowed not only on 25kv electrified routes, but all of the major inter-city ones at that.

It was John Cameron's A4 Pacific No, 60009 *Union of South Africa* that opened up the now-electrified East Coast Main Line for steam, by running out of King's Cross in October 1994. This was exactly 30 years after the same engine had hauled BR's official last steam train from the terminus and was the first steam departure since *Flying Scotsman*'s on August 31, 1969, just prior to its trip to America.

For maximum impact and publicity and now that it was feasible once again, *Scotsman*'s debut in its new guise was to be between King's Cross and York, on the line it had been built for 76 years earlier, and on most of which it had not been seen for 30 years. The great day was July 4, 1999. Whatever may have gone wrong with the sums, this was a triumphant return, beyond what anyone could have imagined. The timescale was tight, the engine was only just completed and run-in on time, and a little paintwork remained unfinished, but its performance was fantastic, and public interest was colossal. An estimated one million people lined the route to see the engine's return — just what the company needed.

Unfortunately, it is probably fair to say that financially at least, it was mostly downhill from then on. When the engine ran, it was like no other, certainly not like any other A3 Pacific, but although most of its trains did sell, there were not enough of them. Privatisation, which had opened up the routes it needed to run on, especially out of the main London termini, also led to the well-publicised backlog of maintenance of the railway system, making it almost impossible to plan tours because large chunks of the system, while theoretically open to all, were in fact often closed to all, for maintenance.

A major problem became the East Coast Main Line as *Flying Scotsman*'s route, King's Cross-to- York, was just too popular with the travelling public, and an extra train travelling at half the speeds of the GNER expresses, simply could not be fitted in at the time it needed to run. Wealthy clientele, having paid several hundred pounds for a ticket to ride behind the world's most famous steam locomotive, did not want to sit in Retford down goods loop for an hour, waiting for a slight lull in the traffic.

The plan to overhaul A4 Pacific No. 60019 *Bittern*, also acquired by Marchington, was abandoned and the engine sold. Marchington had also bought back what was left of *Flying Scotsman*'s second tender, but the project to rebuild this also fell by the wayside and it was sold to the A1 Trust, which was building a brand-new LNER Peppercorn A1 Pacific.

A problem found with *Flying Scotsman* during its first winter of main line operation with a double chimney was that of drifting smoke obscuring the driver's vision. Just as BR had fitted smoke deflectors to cure this problem (eventually), the decision was made to refit them towards the end of 1999. The engine was now very definitely 'non-authentic' in appearance, carrying a 1958-style double chimney and 1962-style smoke deflectors, but turned out in pre-1939 LNER apple green livery.

Other organisers hired *Flying Scotsman* and it found itself on routes it had never visited before – one promoter even ran it from St Pancras to Inverness – but they were soon disenchanted by the nightmare of dealing with the bureaucracy of the privatised railway system. Two things kept the engine going: the unerring commitment of Roland Kennington and his team, and a potentially lucrative contract to haul the 'Orient Express'. The Venice Simplon Orient-Express had been launched in 1984 as a luxury train running a couple of times a week from London to Venice. The train ran in two parts, Victoria to Folkestone and Boulogne to Venice or other European destinations – using superbly restored vintage Pullman cars in Britain and Wagon-Lits dining and sleeping cars on the Continent. Passengers used the ferry to cross the Channel.

Gradually an increasing programme of domestic tours was introduced so that the public could experience travel on Britain's most luxurious train, not cheaply, but without the expense of going to Venice and flying back. There had always been thoughts that steam haulage would be a considerable added attraction on the domestic runs, but the fact that the train was air-braked only, to be compatible with the modern railway system, while steam engines are traditionally vacuum-braked only, made it impracticable until 1995.

The SR Merchant Navy Pacific No. 35028 *Clan Line* was the first preserved main line steam engine to be fitted with air brakes and it proved to be a big hit on the VSOE as the operators and passengers liked having steam on the front of the train. *Flying Scotsman*, also now fitted with air brakes, managed to pick up a piece of the VSOE action on an occasional basis at first, from August 2000, and then as the preferred engine when *Clan Line* retired for its next seven-year overhaul.

VSOE runs generally started from London Victoria, and ran to destinations such as Southampton, Bath, Worcester or Stratford-upon-Avon, although occasionally No. 4472 ran from King's Cross to York or vice versa. VSOE also introduced the 'Northern Belle', offering a similar standard of luxury, albeit in less interesting coaches, for customers in northern cities. *Scotsman* took its turn on this train, sometimes to Scarborough, but often between York and Newcastle.

Tony Marchington had always intended to launch a public limited company so that individuals could buy shares in the engine. That way it would not be owned by an individual, and its future well-being would be assured. A VSOE contract with a guaranteed income from 30 to 40 trains per year was what the company needed to make it attractive to investors.

Flying Scotsman plc was launched on Ofex, the junior stock exchange, on December 3, 2001. Individuals could now buy a share in the *Flying Scotsman* name, and initial interest from the public was described as massive, with eventually more than £1.3 million being raised. One investor who bought a small stake in the engine was none other than Alan Pegler, who had saved it from being scrapped in 1963.

But *Flying Scotsman* still had major problems. Tony Marchington antagonised VSOE management by taking his engine to a traction engine rally in Derbyshire, making it unavailable for several of its contracted jobs. He had first taken it to the Hartington Moor rally in 1999 after a bet in a pub. There were times when there was simply insufficient money for simple running repairs to get *Scotsman* fit for its next VSOE duty, and the A3 was also a little oversized for trips in Kent, a frequent destination for VSOE runs.

Sea Containers Ltd, owners of the VSOE, looked forward to the return of *Clan Line*, and also signed up a couple of other air-braked

Bulleid Pacifics to cover for the A3. It is probably also fair to say that, to make money, *Flying Scotsman* had to make more runs per year on VSOE trips than was really practicable. Nevertheless, when it was good, it was very good; the VSOE is a very heavy train and the A3 worked harder than it ever had in its previous 80 years' service, turning out power beyond anything even Gresley could have dreamed of.

Impressive though its power output was, it actually added little to *Flying Scotsman*'s fame as far as the great British public was concerned. After the initial flurry of interest shown by the media and the public in 1999, little was heard of the engine on its travels, except for occasional rumours about its financial situation and its long-term future.

Chapter 17

THE DIRECTOR'S TALE

✳✳✳

DAVID WARD and *Flying Scotsman* go back a long way – more than half a century, in fact. A railwayman all his working life, David joined the newly fledged British Railways in 1948, based in York. "*Scotsman* was, of course, a regular visitor then, passing through on East Coast Main Line expresses, and I saw it on numerous occasions," he recalled. "Even then there was something about the locomotive – the name is brilliant, very charismatic."

Its reputation and public perception were then, of course, also enhanced by the fact that there was also this romantic train, 'The Flying Scotsman'. "The public often confused the two, and I suspect still do," David said. As he progressed through the BR ranks, No. 60103 – as it was in BR days – continued plying its trade on East Coast metals until the men in grey suits decided in 1963 that its time had come. The engine was, it seemed, destined for the cutter's merciless torch, leaving behind just memories and a record or two.

Alan Pegler, though, had different ideas, and in what was probably the most publicised public salvation of modern times, he bought the locomotive seemingly from under the scrap dealers' noses. "I have complete admiration for Alan and what he did, but in all honesty the railway historians have not portrayed the complete picture," said David. "It wasn't really a case of either Alan buying or *Scotsman* being scrapped. "There were others in the field trying to find the money to buy the loco- motive, but Alan was a member of the BR Eastern Region board and was able — quite understandably — to pull a few strings."

By this time David was based in Norwich, and No. 4472 — as it had become in preservation — was a regular visitor to that cathedral city, running specials in the twilight of BR steam. When that twilight turned to total darkness in August 1968, Alan was allowed to continue running *Flying Scotsman* on the main line under an agreement with BR. This deal was to run until 1971, and meant No. 4472 was exempt from the ban that applied to every other steam locomotive.

Even early in its preserved life, No. 4472 was blazing trails. At this point David became involved with the locomotive for the first time from a formal operational point of view. "A number of people were nominated by BR to liaise with Alan over the running of his special trains, and I was one of them."

Flying Scotsman continued its high-flying career — visiting both the United States and, some years later, Australia — and so did David, who was appointed Director Special Trains in 1984. In this role he was responsible for all non-timetabled trains, which included Royal, VIP and heritage trains involving steam, diesel or electric traction. Once again No. 4472 was within his remit.

In 1994 David retired after a 46-year railway career, but two years later 'the world's most famous steam locomotive' was once again to become a part of his life — only this time in a big, big way. "I received a call from Roland Kennington, who was chief engineer for *Flying Scotsman*," explained David. "He told me a Dr Tony Marchington had bought the locomotive, and asked if I would be interested in looking after the operational side.

"Roland told me a little about Tony, and said that he had bought the engine to put something back into the community. He also told me Tony wanted a 'Rolls-Royce' overhaul, and that there was no problem with money. "I was interested, so I went to see Tony at the headquarters of his pharmaceutical company in Oxford. I had come from a very different world, from a nationalised industry where money was tight and everything disciplined. I took Tony at his word and felt his intentions were sound, so I agreed to become operations director."

With the overhaul completed in 1999, David's days as a retired pensioner were put on the backburner. He was about to enter a new phase in his business life — albeit one that was unpaid — that must have come as a total shock to someone who had spent his entire career in the warm, polarised and orderly embrace of BR. It is to his credit, though, that he adapted to such an unaccustomed environment with aplomb, showing a talent that probably surprised no one who knew him but a fighting spirit that raised a few eyebrows.

The people with whom he worked within the *Flying Scotsman* circle were an eclectic bunch of individuals who probably had only one thing in common: their lives were about to revolve to varying degrees around an elderly 160-ton lady who demanded constant attention. They included a steam locomotive owner who tended to get his own way whatever that way might be, a chief executive who was a qualified solicitor and former Conservative MP, a chief engineer whose knowledge of the locomotive was probably greater than any other living person, and a PR man whose motor industry background was about as far removed from the straitjacketed world of BR as it was possible to get.

David's disagreements with the chief executive — Peter Butler — and the PR man were legion. He would become frustrated at what he perceived as a lack of knowledge on the subject, irritated by an attitude that to him bordered on excessive laissez-faire, and bewildered when a problem that to him was of vital importance wasn't treated with the same reverence by others within the circle.

He accepted, for example, that there was enormous and boundless public and press interest in the locomotive, but failed to see why linesiders and people waiting at stations — none of whom had contributed in any way to saving the engine or keeping it running — should be allowed a free spectacle. On one run to Lincoln, for example, the PR man had arranged to broadcast live from the train via his mobile telephone to the local radio station, reporting on progress at regular intervals as they approached the city.

Such was the public enthusiasm whipped up by this exercise that a crowd estimated at 10,000 was waiting at the station and by the line as *Scotsman* drew in. The PR chap was beside himself with delight as he saw a sea of excited faces thronging the platform, with not an inch of space unoccupied. To him it was a job well done. David was appalled, genuinely feeling those people had no right to enjoy the sight or get in the way of the passengers who had paid a tidy sum to be on board. It was, he said, "a fiasco".

Despite such disagreements, however, David was the ultimate professional, bringing to the party much-needed expertise and a veritable tome of contacts that played their part in keeping the show on the road. He was, arguably, the man who stitched everything together, and he seemed never to resent waking up to yet another problem. It was an extremely busy period, for Tony Marchington and Peter Butler were determined that No. 4472 should travel far and wide, reaching such geographically diverse towns and cities as Norwich, Yeovil, Plymouth, Shrewsbury, York and Inverness.

As operations director David had his hands full, and insiders will tell you that he never let anyone down personally. The financial jinx that is so woven into the locomotive's preservation history began to rear its debilitating head, however. "Early on, Tony had a vision of running a first-class outfit," said David. "He had bought another locomotive — A4 No 60019 *Bittern* — owned a set of Pullman cars, and was also interested in buying a set of sleeping cars. It could have been a very good business, but unfortunately the capital wasn't there. It went downhill in terms of aspiration and the vision.

"Things were okay for a couple of years, but then I began spending a lot of time trying to keep the operation afloat. Suppliers were on stop, and financially we never knew where we stood." The company had a contract with the Venice Simplon Orient-Express luxury train operation but this, too, had its problems. "I remember that once we had to pull out of providing *Scotsman* for three VSOE trains, which meant a loss of income — and all for a repair that would have cost £2,000. It was all a struggle and very depressing."

In another well-publicised episode, No. 4472 was unavailable to the VSOE because Tony wanted it to appear at a country steam fair he was involved with. "The thing I regretted most was letting the VSOE people down, whatever the reason. They were an excellent company to deal with, and I hated making the call giving them the bad news. My allegiance was always to *Flying Scotsman* and not the people who were running it. What went on was doing the name of the locomotive no good at all."

Knowing so intimately the locomotive's financial travails, it could hardly have been a surprise to him when it was announced early in 2004 that *Flying Scotsman* was for sale. However, as with chief engineer Roland Kennington, with whom he forged a close and rewarding business relationship, he had his misgivings about its purchase by the National Railway Museum.

"I think it was wrong to spend public money when private finance was available. Three private owners kept the locomotive operational for 41 years and 445,000 miles — that is a record to be proud of. When *Flying Scotsman* was handed over to the NRM it was in a far better condition than when Tony Marchington bought it eight years earlier and, indeed, was probably in its best condition since being withdrawn in 1963. You have to give him credit for that."

Chapter 18

FOR SALE

✳✳✳

THERE WAS no solution to *Flying Scotsman's* financial problems — the company made a trading loss of almost £500,000
in 2002. Over £2 million had been spent on purchasing and
overhauling the engine, but much of this was in the form of bank loans.
Marchington was, on paper, a multimillionaire, but his fortune was linked
to the value of his company, Oxford Molecular Group plc.

The management team at Flying Scotsman plc expected to command
big city salaries, and the cost of running the company far exceeded the
income generated by the engine. No. 4472 had a value, but it had to be
estimated at around £2.5 million in order to keep the company solvent,
and no one was actually likely to buy it at that price just to clear the
company's debts and overdrafts.

A solution was to sell the rights to the name *Flying Scotsman*, an asset
registered in the name of the company, to an investor who could use it
for financial gain. A shopping centre called the Flying Scotsman theme
park, steam village or even shopping mall was one particular brainchild,
and it was to be built in the Peak District, or perhaps Doncaster, or even
Edinburgh … The idea was that a consortium building a shopping centre
for billions of pounds could call it the Flying Scotsman Shopping Centre

on payment of a couple of million pounds to Flying Scotsman plc, and the use of the name would add much more than a couple of million to the value of the finished item.

It was never clear whether the engine itself was part of the deal, and at best it seems an over-estimation of the value of the name *Flying Scotsman*, famous though it is.

Marchington's involvement ended in July 2003 with his resignation from the board, and he was in fact, declared bankrupt in October, although not exclusively because of *Flying Scotsman*. The financial problems of Flying Scotsman plc simply compounded the problem of the collapse of the share price of Oxford Molecular Group plc.

If it had not been for Dr Marchington, *Flying Scotsman* may never have been overhauled to main line condition again. The plans for the engine at the outset were perfectly realistic, but the overhaul cost so much that the engine simply could not earn sufficient income to pay the interest on the bank loans taken out to pay for it. Few main line steam engines make sufficient money to cover the cost of a seven-year overhaul, and many are paid for with help from the Heritage Lottery Fund.

On November 3, 2003, Ofex suspended trading in the shares of Flying Scotsman plc. With No. 4472 still making regular VSOE appearances, the situation degenerated into farce when the engine was advertised for sale in 2003 by a used car dealer. True, he was a dealer in very expensive vintage vehicles, even if his premises resembled a barn used for rearing hens. An end to this chapter in *Flying Scotsman's* history was clearly approaching, but what might happen to it now?

SAVE OUR SCOTSMAN

With it becoming increasingly inevitable that *Flying Scotsman* would be sold again, considerable public debate ensued. Fortunately, though, it was not all just talk, and serious moves were afoot to raise the money needed to purchase it so that it would remain in Britain and remain in steam.

The management of Flying Scotsman plc still considered that it had an asset worth a considerable sum, and it was determined to see it sold for an amount that would clear all the debts of the company. What was clearly important to the company was to ensure there were several bidders, and if this included overseas collectors, then so be it.

Among the rumour and counter-rumour it was stated that the National Railway Museum would like to acquire it, and not just for display as a static exhibit. Jeremy Hosking, new owner of A4 Pacific No. 60019 *Bittern*, was interested but not at any price. A Midlands-based businessman also emerged as a serious bidder, and there continued to be rumours that overseas investors were interested.

At the time, Flying Scotsman plc was still trading and the engine was still running. The company was in serious debt to its bank, and therefore could not sell the asset for less than its market value, and this could only really be established by offering it for sale by tender. The sale was officially announced on February 16, 2004, with a deadline of April 2 for receipt of sealed bids. Within three days, the National Railway Museum launched its Save Our Scotsman campaign, aimed at raising sufficient money to be assured of obtaining the engine.

This meant it had to be sure of exceeding the bids of unknown potential purchasers, and offering sufficient to satisfy the company's creditors, primarily the bank. It was at this stage that it became apparent just how famous the engine was, and how much the British public cared about it, as £365,000 was raised from public donations within the first five weeks, a level of support unprecedented in the history of steam preservation. It was without doubt one of the most incredible and successful fundraising initiatives ever staged by a state-owned British museum, to acquire a historic artefact for the nation.

The museum's campaign was supported by MPs across the country, companies such as GNER, which then operated trains on 'the route of the *Flying Scotsman*', newspapers such as the *Yorkshire Post*, and perhaps most notably, Sir Richard Branson, head of Virgin Trains, who offered to match pound-for-pound the amount donated by the public.

The museum also made an application to the National Heritage Memorial Fund, the parent body of the Heritage Lottery Fund, for emergency funding, an amount of £1.8 million being agreed. What caused the huge upsurge in public interest and support? Was it the threat of the engine disappearing overseas forever? Or was it the fact that *Flying Scotsman* had simply not existed for almost ten years as far as most people north of London were concerned anyway?

After a period out of use followed by a long-running overhaul, it had hauled expensive upmarket trains, mainly south and west of London in recent years. Perhaps the public were really saying 'we want our engine back', to pull trains for ordinary people in all parts of Britain. Certainly, they did not just say so; they put their money where their mouths were. It was publicly announced on April 19 that the National Railway Museum's bid of £2.31 million had been successful. At least one other bid came close to this figure, but the engine was now in public ownership, finally a part of the National Collection, thanks to the NRM's initiative, public donations, the NHMF and all the individuals and companies that had lent their support.

Now it was safe, a new chapter in the engine's history was about to unfold. The NRM bought the engine to run it, and its arrival at York was timed to coincide with Railfest, a weeklong celebration at the end of May 2004, of the 200th anniversary of Trevithick's Penydarren tramway locomotive, and the centenary of *City of Truro*'s 100mph run in 1904.

No. 4472 would arrive at York hauling a special train from Doncaster conveying invited guests. It was a high-profile media occasion and although the engine ran light from Southall as far as Doncaster, it was found to have a leaking boiler tube on arrival and was unable to haul the train, having to be towed to York and pushed into the exhibition site by a diesel. Not a good start!

SAVED FOR THE NATION

Flying Scotsman had a major rebuild during 1995-99 at a cost of around £800,000. It remained main-line certified throughout the ensuing five years, but covered a relatively low mileage in view of some long periods not being used, mainly because of shortages of money. It had never actually broken down while hauling a train during its period working for Flying Scotsman Railways, mainly on Orient Express duties. It was reasonable to assume therefore, that it was in fairly good condition and with two years left on its main line boiler certificate, the National Railway Museum decided to keep it running after the purchase, and not to carry out a full overhaul until the certificate expired in 2006.

The public had supported the appeal and it was only reasonable to make the engine available for the public to ride behind as quickly and as often as possible. To have immediately stripped it down and for it not to be on public view for a couple of years was not really an option. There was no question of letting it loose on ambitious main line runs, or hiring it to any heritage lines, but although the engine could not steam when it arrived at York, it was not a major repair job. The museum was confident in advertising trains from York to Scarborough during the summer, to be hauled by its famous new acquisition. Interest in these trains, running twice a day, three times a week at just £25 for the 84-mile round-trip, was enormous, with 1,000 tickets sold on the first day they were on sale.

On the first day of service, July 20, there were some shunting problems in actually getting the engine and its train into the platform ready for departure, and *Scotsman* was over an hour late. As it pulled in to platform 5, though, it was greeted by a spontaneous round of applause. When did any other steam engine ever earn that?

Unfortunately, *Flying Scotsman* broke down several times during the summer, often with the eyes of the media watching it. They were relatively minor problems and quite quickly repaired, but disappointing for the new owners and for the travelling public.

The price paid for the engine in 2004, although a lot of money, left enough in the kitty to finance the engine's ongoing maintenance, and the Scarborough runs made a small profit. A cash injection from Yorkshire Forward, for a permanent exhibition at York based around the engine, and to assist in its mechanical upkeep, also helped to keep No. 4472's bank balance healthy. It was basically considered fit to run at the end of the 2004 summer, and not in need of an immediate major overhaul.

However, there were a number of fairly major jobs required, to ensure a more reliable level of performance in the summer of 2005. The museum has a policy of carrying out maintenance not only in-house, but in full view of the public. Despite being in a fairly heavily dismantled state for the next nine months, No. 4472 remained on view and was a major attraction in the museum.

Unfortunately, it missed another high-profile run in September when it had been scheduled to haul a train conveying the then prime minister, Tony Blair, to the opening of Locomotion, the National Railway Museum at Shildon, in the PM's Sedgefield constituency.

Jim Rees, then rail vehicle collections manager at the NRM, explains: "When the engine arrived, it could best be described as tired. Yes, it had been putting up terrific performances hauling loads beyond what was expected during its BR service, but it is an A3, fitted with an A4 boiler running at A4 pressure of 250psi. Gresley had not designed the mechanics of the A3 to handle this kind of power and it was having an effect on the engine. There was a crack in one of the cylinders that had to be repaired, and the engine needed a complete piston and valve examination and overhaul. Much work also centred on the boiler."

The timescale for the work on No. 4472 during the winter was very tight, but the engine was ready for the 2005 season. The public had clearly forgiven its indiscretions the previous summer and advance ticket sales were phenomenal. For the first couple of days in Spring Bank Holiday week, it delighted its public with its runs to the seaside.

Volunteer supporters working on the train reported a carnival atmosphere on board, such was the relief that No. 4472 was finally back in tip-top condition. Or was it? Even in the first week, problems arose with the boiler, and *Scotsman* failed to appear on the third day. It was repaired in time for the start of the main season though.

There were more problems and 'Hogwarts Castle' stood in for a week, but the passengers were not impressed. There were reports of people travelling hundreds or thousands of miles for the ride, and bursting into tears when *Scotsman* failed to show. The museum's Gresley V2 2-6-2, No. 60800 *Green Arrow* also took a turn, but after more boiler repairs, *Flying Scotsman* was soon back in business. But what should be a routine steam trip to the seaside was turning into a nightmare for everyone concerned for the second year running, and it was starting to become the most famous steam engine in the world for all the wrong reasons.

The run from York to Scarborough is one of the easiest jobs a large main line steam engine can be asked to do, yet after a lifetime of breaking speed, endurance and haulage records with comparatively few mechanical problems, the A3 could not be relied on to get there and back.

The engine was 82 years old — the second-oldest running on the main line in 2005. While other main line steam engines were making on average a dozen runs a year, such was the public demand for the most famous one of all, that *Flying Scotsman* was expected to run three days a week for a three-month period — and that is far harder work, for the engine and its operators, than was expected of any of its younger competitors.

Flying Scotsman did travel to Crewe to take part in The Great Gathering in September and then moved on to the Tyseley Locomotive Works in Birmingham, from where it hauled a farewell series of dining trains to Didcot or Trent Junction before Christmas. It took a couple of coaches back to Carnforth afterwards and returned to the NRM at York where its fire was dropped for the last time on December 20, ready for its overhaul to start.

Chapter 19

BACK TO THE
FUTURE

✳✳✳

BRIAN SHARPE met Jim Rees in 2005, then rail vehicle collections manager at the National Railway Museum at York.

The National Railway Museum completed its historic purchase of *Flying Scotsman* on behalf of the nation in the spring of 2004, and Jim Rees assumed responsibility for the ongoing preservation and operation of the engine. Just as it was important to clarify why Alan Pegler had bought it in 1963, so it seemed appropriate to understand why the latest (and last) purchaser had bought it.

The answer was quite simple really: "This is the largest railway museum in the world and *Flying Scotsman* is the world's most famous steam engine." But Jim Rees added: "It became clear during the fundraising campaign that most people, both the public and the media, were amazed to find that the museum did not in fact, already own it." Jim was never under any illusions that it was going to be easy, and it was a hard slog from the outset.

Just like Alan Pegler more than 40 years earlier, the museum did not just buy the engine to preserve it, or to stop it being scrapped or sold abroad. The engine was the public's favourite, and there was overwhelming demand for it to run. There was absolutely no thought of the engine being 'stuffed and mounted' in the foreseeable future. He said: "We bought it to keep it running. The engine was mechanically very tired when it arrived, but the main problems were with the boiler.

"The first thing we did was to immediately reduce the boiler pressure to 220psi, as the A3s were designed to run at that. Another early problem to address was that the engine did not own a chimney. When the last A3, No. 60041 *Salmon Trout*, was scrapped in 1966, a chap bought its chimney. Many years later, Roland Kennington had found out about it, borrowed it and put it on the engine when it had its rebuild to double chimney condition in 1993. But it never actually changed hands, so we had to negotiate to buy it from the chap who owned it." Jim added: "People think we could just put the old single chimney back on; they don't realise that there is a big difference inside the smokebox between a double-chimney A3 and a single-chimney one."

Had the engine made money since the museum bought it? "There has been a small surplus on the operation so far, despite us having had to carry out more running maintenance and repairs than had been hoped for." Again, like Alan Pegler so many years earlier, the museum did not expect to make money out of the engine. The problem over the first two summers was to balance expenditure on essential repairs to keep the engine going and investment in its long-term future.

It had two boilers. The one it carried in 2005 was an A4 one, last carried by No. 60019 *Bittern* in 1965, which was bought by Alan Pegler as a spare. It was fitted to the engine at an overhaul in 1978. The other was an A3 boiler that was carried by A3 No. 60041 *Salmon Trout*, but removed in 1963, three years before it was scrapped. This boiler was fitted to No. 4472 when it was overhauled at Darlington in 1964, 12 months after Pegler had bought it. The A3 boiler, not used since 1978, was the better of the two, and had already been sent to the works of Riley & Son Engineering at Bury "for a major rebuild, not just an overhaul, to make it fit for the next 20 years at least".

This would speed up the engine's return to service, and make it more of an authentic A3. "But our problem," said Jim, "is that, having decided that the present boiler is never likely to see further service, we have not wanted to spend money on repairing it during the last two years, other than necessary running maintenance." Coincidentally, Alan Pegler bought the cylinders from *Salmon Trout* when it was scrapped. The middle one was fitted during the Marchington overhaul. The museum was going to fit one of the outside ones, so *Flying Scotsman* was going to run with *Salmon Trout*'s one-time boiler, and two of its cylinders.

How much would the overhaul cost? Jim's estimate was £600,000. He knew the engine well by then and did not expect any surprises. The money was available, boosted by a Lottery grant, and a recent and very welcome donation of £75,000 from GNER, which advertised its East Coast Main Line expresses as running on 'The Route of the Flying Scotsman'. The boiler had gone to Bury already.

The target for completion was August 2007, which was a very tight timescale, especially as the work would be done by the museum's engineering staff of just three. Some other work as well as the boiler would be subcontracted out. The museum readily admitted that it wanted a more reliable engine. The aim was to have an engine in August 2007 "which will do everything an A3 should do, nothing more and nothing less. It has no records to break and nothing to prove".

There was one obvious and so-far-unanswered question. Was it going to be a single-chimneyed or double-chimneyed A3? "The jury is still out on that one," said Jim. "The museum bought the engine to run it, so performance and reliability is an important issue, perhaps uniquely in the case of this one engine in the collection. It does not need to haul the heavy trains it hauled in its VSOE days, and the A3s were not built to climb hills, but the double chimney and Kylchap transformed the engine's performance in the 1950s."

Flying Scotsman ran far more economically in its second season at York, but it also surprised everyone by continuing to steam one day even when a superheater element came apart from the header. This was a major problem, yet the engine carried on, something it probably could not have done in single-chimney form. With a desire to run the engine, not just to Scarborough, but also over the Settle & Carlisle line, and hopefully even from King's Cross to Edinburgh occasionally, the advantage of the double chimney could be disregarded.

There is a strong argument that a museum is responsible for conserving an item in exactly the condition in which it obtained it, but perhaps the strongest argument was the simple fact that there simply was not an A3 single chimney and blastpipe available to fit to the engine. Nevertheless, there were also compelling arguments for reverting to the single chimney, as this was the form that the public recognised. Because many of its followers in Yorkshire had not seen the engine for so long, many believed that it was the museum that had fitted the smoke deflectors, and they wanted them removed.

Jim actually tried running it to Scarborough without deflectors, but admitted to being surprised as to what difference it made. Even in summer, and only running at above 40mph for parts of the journey, the lack of smoke deflectors caused visibility problems on a double-chimneyed engine. Even so, it was thought that it may be possible to run occasionally without them, if only on the Scarborough run. So a decision had yet to be made on the issue of the chimney, and time was running out. The museum's collections committee was going to make the decision in the fairly near future.

Jim pointed out one feature that would definitely please the purists though, and pointed to the corner of his office. "There is a proper, fully working A3 whistle in that cupboard." The whistle carried by No. 4472 was wearing out, not really loud enough, and was replaced by Roland Kennington with a much louder chime whistle a few years earlier.

The engine would remain on view during its overhaul, but there was also going to be a permanent 'Flying Scotsman' exhibition at the museum. This would relate to the train and, while No. 4472 might occasionally be part of the exhibition, it would be equally valid for the Stirling Single or a Deltic diesel to feature, to clear up that continuing confusion between *Flying Scotsman* the engine and 'The Flying Scotsman' the train.

"*Flying Scotsman* is the most famous steam engine in the world, and it is still continuing to become more famous," said Jim. However, he felt that further overseas exploits were unlikely, and had not actually contributed much to the engine's following in Britain. The engine had undoubtedly diverted management and staff time and effort since the museum bought it, but he was adamant that the engine would continue to run, because the management and staff at the museum were absolutely committed to it. "They all *want* to keep it going!" Long may the legend steam on!

THE OVERHAUL

To say that the overhaul did not go according to plan would be a bit of an understatement.

The National Railway Museum has excellent engineering facilities and a good record for carrying out overhauls of big steam locomotives to main line standards, but *Flying Scotsman* was different.

The museum had returned LMS Princess Coronation Pacific No. 46229 *Duchess of Hamilton* to steam in 1980 and LNER A4 Pacific No. 4468 *Mallard* to steam in 1986 but these were straightforward, if quite expensive overhauls, and there were people involved from railway backgrounds with skills learnt in the steam age.

It was now 30 years since *Mallard*'s overhaul and the museum was now a very different place, having had little involvement in running steam engines on the main line for some time. Not only was *Flying Scotsman* in extremely rundown condition but also much time and money would have to be spent in undoing some of what had been done to the engine, as well as repairing and restoring it.

The museum had had considerable involvement in running steam trains in the 1970s, 1980s and 1990s but circumstances had changed somewhat in that its priorities were becoming more clearly defined and conserving historic artefacts was now taking precedence over running steam trains. There was always a desire in some quarters to return *Flying Scotsman* to as close to authentic A3 condition as possible. For example, *Scotsman* had two boilers, both in poor condition, and the decision was made to use the slightly better one of the two as it was a genuine A3 boiler, as opposed to the rather worse one carried by the engine on arrival, which had been designed for an A4.

There had been three different heads of the museum in the previous 11 years, each with markedly different management styles. The staff involved with the overhaul, particularly the chief engineers, changed at regular intervals and there was a significant change in the management structure at one point. Ongoing discussion, if not disagreement, continued over whether the priority was to recreate as far as possible a genuine A3 Pacific or to produce the most efficient and reliable engine possible. There was also a desire to keep the engine on show to the public while the overhaul progressed. Initially, the museum had bought not so much a steam engine, but a brand name with a huge public following but gradually, as the overhaul dragged on, the museum's thinking predominated more and historical accuracy became more important.

It was going to be a huge undertaking but when the overhaul commenced in January 2006, with a target date for completion of August 2007, the A3 spare boiler had been sent to Riley & Son Engineering in Bury, a company with a proven track record of carrying out overhauls of steam engines to main line standards, and also of operating the engines in all parts of the country. A new firebox was needed, which would delay the return to service by a year and, in fact, problems in its construction led to a further 12 months' delay. Consideration had been given to using the A4 boiler after all or even having a new one built in Germany, but the overriding desire was still to restore the A3 with a genuine A3 boiler.

In 2009 a public appeal was launched for a further £250,000 to complete the job. Also in 2009, Andrew Scott, who had spearheaded the Save Our Scotsman campaign in 2004, retired and Steve Davies was appointed as the NRM director.

Chris Beet was appointed as engineering and operations manager and the aim was to complete the locomotive's 'bottom end' ready to be moved to Rileys at Bury in 2010 for the boiler to be fitted, with a view to entering service in 2011. In fact, the frames moved to and fro across the Pennines by road more than once.

However, by 2011 reassembly was approaching completion with the boiler test successfully completed and it was reunited with the frames on March 25 with a test run scheduled for June. A decision had been made to paint it in wartime black livery, numbered 502 on one cabside and 103 on the other, but it would quickly revert to apple green once back in traffic.

Accordingly, the engine was moved by road from Rileys at Bury to the NRM on May 27 and put on show to the public. However, when it returned to Bury, major problems with the locomotive's frames were discovered, almost by chance, and these turned out to be even more serious than was feared.

Major faults were found on more than one occasion, either dating back to previous overhauls or in things overlooked during this latest one.

The engine had to be dismantled again, at Bury this time, and major and expensive remedial work was carried out. It is clear that these ongoing problems led to the departure of more than one of the NRM engineers in charge at various times.

It got to a point where the museum even considered abandoning its plans to return the engine to steam and Bob Meanley from Tyseley Locomotive Works and Professor Roger Kemp of Lancaster University were engaged by Steve Davies, to review progress and make recommendations as to how best to proceed. The lengthy report produced in October 2012 did at least vindicate the museum's stance that the engine was simply worn out when it was acquired but also that it should have carried out a far more detailed investigation into its condition before purchase, particularly ultrasonic testing of the frames.

A further report was then commissioned from First Class Partnerships as to how to now proceed to complete the project. This found that the museum no longer had the resources to manage a project of this magnitude. There had been inadequate supervision of subcontractors and parts were simply disappearing. However, *Scotsman* also appeared at the June 2012 NRM Railfest, still in black and Steve Davies was still confident it would run by the end of the year.

A second report was carried out by First Class Partnerships, which recommended that the museum should seek bids for the completion of the overhaul and the ongoing operation of *Flying Scotsman*. Much of the work carried out, including the initial boiler overhaul, had already been subcontracted to Riley & Son Engineering in Bury.

The successful bidder, not surprisingly, was Rileys and a contract was signed in October 2013. The boiler was lifted from the frames and the frames were returned to Riley's works at Bury yet again. The contract also stated that the company would have exclusive responsibility for maintaining and operating the engine for two years after its return to steam.

In these late stages of the overhaul it was discovered at various times that the frames were out of true; there were cracks around the hornguides that hold the axleboxes; the main frame stretchers and the centre cylinder motion bracket were cracked; the middle cylinder was out of line and the holes in the frames to which the cylinders were bolted, had become oval, meaning replacement of the front sections of frame.

It was felt that many of these problems had been caused by the engine being worked harder than it was designed for, particularly during its period working heavy VSOE trains in the south of England, with a higher boiler pressure and larger cylinders than in the original A3 specification.

In 2013, Paul Kirkman took over from Steve Davies as NRM director. The overhaul and additional remedial work progressed steadily at Bury and there was now light at the end of the tunnel.

Flying Scotsman was away from Britain, let alone its main lines, for three years from 1969 to 1972 but away from its home ground, the East Coast Main Line, for a further 11 years, largely as a result of BR steam operating politics. Having made just three runs and then only north of Peterborough in 1983, it did not return to the ECML again until 1999 – a further 16-year absence, and did not run on the main line at all for seven years from 1992. So, despite the ten years now elapsed since it last ran on Britain's main lines, it is worth remembering that this was not the first time it had spent extended periods away from its native ECML or even any British main line.

THE RETURN OF THE UNTHINKABLE

The most obvious example of where conservation now appears to take priority over commercialism is in the appearance of the locomotive today. The public has generally recognised *Flying Scotsman* as an LNER apple-green locomotive carrying the number 4472. As we saw earlier, in BR days it carried the number 60103 and in 1958 it was fitted with a double chimney, which improved its performance significantly. Alan Pegler, who bought it in 1963, had the single chimney refitted and it reverted to its LNER livery and number – the condition in which it achieved its fame.

A double chimney was fitted again in 1994 and to be authentic, the BR livery and number were reapplied, although it did no main line running in this condition. When Dr Marchington bought it, the double chimney was retained but it reverted to LNER livery, which was totally inaccurate but it retained most of what the public identified with visually while being able to produce the better performance necessary to handle the size of trains it found itself dealing with.

The National Railway Museum bought the engine to run it, and in one respect museum thinking was overruled largely on cost grounds because as well as impairing the engine's performance, refitting a single chimney would have been prohibitively expensive. However, the museum had also bought a famous brand name and part of the brand that the public identifies with is the apple-green colour and the number 4472. It was always considered unthinkable that the engine would not emerge as No. 4472 in apple green, whether with a single or double chimney.

Museum thinking has held sway in the end though. It has a double chimney, it has smoke deflectors; therefore, to be historically accurate it must be painted Brunswick green and carry the number 60103. Some doubted that the public would take to this but the engine had been out of the public eye for so long that many have actually forgotten what *Flying Scotsman* looked like. Certainly, the early indications from the first runs, in wartime black and not even carrying its nameplates, were that the public was just glad to see it back.

When Flying Scotsman entered the NRM workshops for its much-needed overhaul, a brand-new LNER Pacific was nearing completion just down the road at Darlington. It is ironic that in *Scotsman's* absence, new A1 Pacific No. 60163 *Tornado* probably assumed the A3's mantle as Britain's most famous express steam engine, painted in apple-green livery.

However, *Flying Scotsman* has always been rather more chameleon-like than most express engines and changes in its appearance have been regular and, at times, quite drastic over the past 93 years. It was purchased by Alan Pegler in January 1963 for active service on the main line. Alan bought it largely as he remembered seeing it at the Wembley exhibition in 1924 and the apple-green livery had left an impression on him.

It was not practicable in 1963 to return the engine to original Gresley A1 specification, as in 1924, but Pegler went as far as he could. Doncaster works returned his engine to single-chimney form and painted it in LNER apple green as No. 4472, but it was still an A3, not an A1 and the only time it has run in preservation in historically accurate condition until now was its short period in BR green livery in 1994/5 when it only ran on heritage lines.

On December 20, 2015, the fire was lit in the firebox of the newly overhauled A3 at Bury, ten years to the day since the fire was last dropped at the NRM at York...

Chapter 20

THE LEGEND AWAKENS

* * *

O N A cold Wednesday evening, January 6, 2016, *Flying Scotsman* eased out of its Bury restoration base and undertook its first test runs along the East Lancashire Railway's Heywood extension.

Its first run in ten years replaced a planned press launch by the National Railway Museum, which was postponed for another two days while what appeared to be a minor problem with the Gresley A3's air pump was attended to.

The test runs followed the lighting of the first fire in the overhauled locomotive's firebox inside Riley & Son (E) Ltd's engineering works shortly before midnight on December 20, ten years to the days since the last time its fire was dropped.

The runs marked another milestone in preservation and the first page of the latest chapter in the illustrious history of the A3 which, back in 1934, became the first steam locomotive in the world to officially break the 100mph barrier.

At the rearranged press launch at Bury (Bolton Street) station on January 8, Noel Hartley, the National Railway Museum's rail operations manager, discussed the commissioning process and future plans for the locomotive

Ian Hewitt from Blackburn-based Heritage Painting spoke about the work needed to transform the locomotive from wartime black livery into BR green livery in the Bury workshops in late January and early February.

Ian Riley from Bury, expressed his pride in having restored *Flying Scotsman* to steam.

Colin Green, co-director at Riley & Son, which was appointed in October 2013 to complete the work on the locomotive and manage and maintain it during its first two years of operation, said: "These are the first stages of bringing it back to the main line and despite being self-confessed men of iron we're really quite emotional to see it move under its own steam at last after years of hard work."

New ELR chairman Mike Kelly said "We're so honoured to host this iconic steam locomotive first in its big return year."

Museum director Paul Kirkman said: "Along with all our generous supporters for this complex project to bring a 1920s-built cultural icon back to life, we have all been looking forward to the day when *Flying Scotsman* is once again running on Britain's tracks. Even though we still have the rest of the commissioning phase to get through, including the main line test runs, we are so thrilled this historic day has finally come to pass."

Heritage Minister Tracey Crouch said: "This is a wonderful way to tell the story of this iconic and well-travelled locomotive and will ensure that people now and in the future understand why it is such an important part of Great Britain's heritage."

The press launch was followed by two weekends in January, of No. 60103 hauling passenger trains on the ELR as part of the commissioning process. Unfortunately, the engine chose not to cooperate. It was steam tight, ran like a sewing machine and sounded spot-on even in reverse, but stopping was another matter. Despite strenuous efforts by the restoration team, the compressor which works the air brakes would still not work properly and another engine had to provide the air on the first weekend of public runs on the ELR. In fact, the solution to this was to use another compressor borrowed from another engine.

Some years earlier, the museum had announced that its operating partner would be West Coast Railways of Carnforth. The engine's main line railtour programme would however be promoted by two companies: Steam Dreams and the Railway Touring Company. Accordingly, both companies announced programmes of quite intensive operations by *Flying Scotsman* in 2016, with West Coast as Train Operating Company.

No. 60103 was set for its first main line test run from West Coast Railways' base at Carnforth on January 20 which was to be followed by its railtour debut for the Railway Touring Company from Manchester to Carlisle via the Settle & Carlisle route, returning via Shap on the 23rd. For its public trains on the ELR and the first main line railtour, the engine still carried wartime black livery. Although the newly fitted compressor functioned properly, the necessary paperwork could not be completed in time for a main line test run before January 23 so this was postponed, resulting in *Flying Scotsman* being unable to meet its commitment for the first railtour.

However, agreement was reached for *Flying Scotsman* to head another 'Winter Cumbrian Mountain Express' for the RTC scheduled for February 6, and the engine was towed to Carnforth on January 27 to be prepared for a main line test run. It had been no less than 24 years since *Flying Scotsman* had been seen at what had been its home base for many years.

On Thursday, February 4, the engine ran from Carnforth to Hellifield and back with its support coach. On arrival back at Carnforth, the A3 coupled on to a nine-coach set of stock and with a Class 47 diesel on the rear, set off once again for Hellifield, 21 minutes late after some delays in shunting.

Arrival at Hellifield was a couple of minutes early and the engine was running well at main line speeds. Unprecedented crowds lined the route and there was even a police presence on Hellifield station. After taking water, Scotsman trundled down the Ribble Valley and after pausing at Whalley, took its heavy train up the 1-in-68 gradient through Langho to Wilpshire with a minimum of fuss.

Planned stops at Blackburn and Preston were severely curtailed and after a fast run along the West Coast Main Line through Lancaster, arrival back at Carnforth was two hours early. The engine had performed impeccably and was pronounced fit to haul the public railtour two days later.

On Saturday, February 6, the train arrived at Carnforth electrically hauled but there were delays in changing engines and *Flying Scotsman* steamed out of the loop 17 minutes late. With a Virgin Trains Pendolino on its tail, the tour was looped at Grayrigg for the express to overtake and was 31 minutes late passing Tebay and starting the climb to Shap summit. There was a diesel on the rear of its train which did help to an extent but nevertheless, *Scotsman* was holding 67mph at Tebay and topped the summit at 38mph. Downhill towards Carlisle, the A3 roared through Penrith station at 72mph and pulled into the border city in just over 82 minutes from Carnforth to be greeted by a huge crowd of well-wishers who had braved the rain.

Departure from Citadel station was 11 down but Scotsman was lucky to have been allowed to use the Settle & Carlisle line as it had been closed by a landslip the previous day and the train had to use the down line from Howe & Co.'s sidings to Culgaith. After a quick water stop at Appleby and with the diesel still giving some assistance, *Scotsman* topped the 1-in-100 to Ais Gill summit in 24 minutes from Appleby at 48mph.

Despite the weather and almost in darkness, there were crowds at Ribblehead reminiscent of those on August 11, 1968 witnessing BR's last-ever steam train. The A3 left its train at Farington Junction near Preston and returned to Carnforth, but only briefly as it was soon on its way to York where it arrived at around midnight, ready for preparations for the next trip. Unfortunately, a driving wheel bearing was found to have run slightly warm.

The leading pair of driving wheels were removed and sent to Bury with the bearings to be examined and the problem rectified while the engine was being painted at York. The wheels were refitted on Saturday, February 19 and everything reassembled ready for a test run to Scarborough on the Monday. However, by then the bombshell had struck as West Coast Railways had advised that it would be unable to operate the inaugural run from King's Cross to York as planned.

This was not entirely unexpected and moves had already been made to have *Scotsman* registered by the only other company able to run steam engines on the national network, D.B. Schenker, essentially a freight haulage company but with steam drivers on its books. Paperwork was quickly finalised and DBS took over not only the test run but the move to London and the prestige official relaunch train from King's Cross to York on Thursday 25th.

Meanwhile the job of repainting the engine continued at York. For its official launch in February 2016, the engine appeared in fully authentic BR Brunswick green livery as No. 60103 still with its double chimney and smoke deflectors. It had even had its smokebox handrail split and the front numberplate correctly positioned on the top hinge, to make it look exactly as when it was withdrawn by BR in January 1963.

Ian Hewitt of Heritage Painting said part way through the job: "We've painted Gresley greats before including the world's fastest locomotive, *Mallard*, but it's an absolute honour to transform *Flying Scotsman*, the most famous of them all into its new livery. BR green will be recognised by many who saw the engine under public ownership in the late fifties and early sixties, and it will be accompanied by black and orange lining and the BR crest.

"It's taken us just over a week to get to this point, and its used 45 litres of paint plus 20 litres of varnish, 20 litres of thinners and hours of painstaking work for our team of five."

It never rains but it pours and during the final fitness-to-run examination on Sunday 20th, a cracked driving wheel spring was discovered. Welding repairs at Rotherham delayed the test run to Scarborough until Tuesday 23rd. Departure from York was four minutes late but arrival at Scarborough was five minutes early.

After a successful run, the engine set off from York on the morning of Tuesday 24th to run to Wembley where it spent the night.

The last-minute problems were quickly forgotten as they evaporated in the bright sunshine which graced many parts of the East Coast Main Line on Thursday, February 25, as *Flying Scotsman* triumphantly headed its comeback train from King's Cross to York.

The trip marked the start of the National Railway Museum's Scotsman season, sponsored by then ECML operator Virgin Trains, whose chairman Richard Branson stumped up a significant portion of the £2.31 million needed to buy the A3 from its previous owner in 2004.

The media interest was of course intense at King's Cross as *Flying Scotsman* prepared to depart at 7.40am. The DBS driver was Paul Major with fireman Dave Proctor and traction inspector Sean Levell. The 11-coach train weighed 391 tons tare, probably about 420 tons full, a good test for the engine.

There was an air of expectancy as the train departed on time and plunged into the darkness of Gasworks tunnel. The early stages of the run were taken very slowly on a fine and dry but very cold morning and it was not until after Hitchin that the engine was really allowed to run. Steam engines are strictly limited to a maximum speed of 75mph these days but by Three Counties the A3 was up to 72½ mph, reaching a maximum of 74½ mph after Arlesey and maintaining speed at this level downhill past Biggleswade, Sandy and Tempsford. *Flying Scotsman* is clearly a very free-running engine aided by the long travel valves and 6ft 8in driving wheels that make it a true racehorse.

Unfortunately, this fine running was short-lived and the train was brought to a sudden stop near St Neots as members of the public were standing close to the track. This delayed not only *Scotsman* but other expresses and the special was 21 minutes late at its first water stop at Holme Fen and arrived at Peterborough for a crew change over 27 minutes late.

From Peterborough, the driver was Steve Hanczar in charge with fireman Jim Clarke and traction inspector Jim Smith. The climb of Stoke Bank was taken slowly but there was another spell of 75mph running after Grantham past Barkston and Hougham until a stop in Claypole loop for an express to overtake. Having gained a bit of time, the stop at Retford was longer than expected and departure was at midday, still 25 minutes late, behind the 10.30am King's Cross to Newcastle, but in front of the 10.35 King's Cross to Leeds

With speed in the 70s north of Doncaster, *Scotsman* was regaining lost time again but then lineside trespass once again intervened, causing another 27 minutes of delay and the train drew into platform 9 at York to be met by huge crowds, just over 53 minutes late.

A total of 297 VIPs, fundraisers, competition winners and members of the public who had paid up to £450 each were on board the trip.

Former cabinet minister Michael Portillo said that he was very excited to be travelling on the train prior to filming on and around the locomotive at York filming for a forthcoming episode of his BBC documentary series *Great British Railway Journeys*. He praised the A3's designer Sir Nigel Gresley for having "an eye for engineering, for design, for style and for marketing".

Amongst the VIP guests was Ron Kennedy, 83, who started his career as a teenage cleaner at King's Cross and who drove *Flying Scotsman* from 1956 until it was sold to Ffestiniog Railway saviour Alan Pegler in 1963.

Also on board was former owner and multimillionaire enthusiast Sir William McAlpine and his wife Judy. "It's a wonderful locomotive like a beautiful woman," he said. "She's in the right place doing the right thing and very much loved by everybody, and the wonderful thing about her, she makes people smile, people love her."

Huge crowds thronged the lineside and far too many thought it acceptable to climb the railway fences and actually stand on the track to get a better view and twice the train and other ECML expresses had to be brought to a stand as such people were endangering their lives by such behaviour. Despite delays caused by these unscheduled stops, it was a triumphant return, and although some passengers on other services were delayed, most were happy to have seen *Scotsman* on its return to the ECML after so many years.

After being literally embraced by swelling crowds on platform 9 at York, *Flying Scotsman* was uncoupled from its train and steamed into the museum's north yard, for a final welcome-home ceremony and speeches to mark the start of Scotsman season.

NRM director Paul Kirkman said: "We have all been looking forward to the day when *Flying Scotsman* steams home to York along the East Coast Main line and now this historic moment has finally come to pass. This celebratory journey marks a new stage in this steam icon's long and colourful history, and is a tribute to all the people who have worked so hard to make this happen, from those that have worked on the restoration itself to the public that donated to our appeal to bring this legend back to life."

David Horne, managing director of Virgin Trains on its East Coast route, said: "*Flying Scotsman* has an incredible history and we're proud to be sponsoring a season celebrating its return to the tracks which starts with today's inaugural run. In the pre-war era, the 'Flying Scotsman 'symbolized speed and style — service qualities which remain important to our customers today."

Sir Peter Hendy, chairman of Network Rail, said: "It's great to see this magnificent symbol of Britain's railway heritage and technology once again running on our tracks. Since this engine was making its regular trips from London to Edinburgh the journey time has halved, frequency quadrupled and the levels of service and comfort are incomparable. Today, as then, the railways are playing a vital role in economic growth, creating jobs and building homes in Britain. Alongside celebrating the glorious history of the oldest railway in the world we also look forward to investing to continue the huge contribution the railway makes to the future of the UK."

The speeches in the yard were given in front of train passengers and invited media, along with Dame Mary Doreen Archer, Lady Archer of Weston-super-Mare, chairman of the trustees of the National Science Museum Group and Sir Peter Luff, chairman of the National Heritage Memorial Fund and Heritage Lottery Fund.

The NRM's 'Starring Scotsman' season examining the locomotive's claims to worldwide fame was opened following its triumphant return to York, and ran until June 19.

From March 25, Stunts, Speed and Style, a free six-week display in the Museum's Great Hall, told the story of the renowned luxury 'Flying Scotsman' service between London and Edinburgh through the eras.

Visitors were able to board the cabs of four different locomotives, including *Flying Scotsman* itself, which hauled the famous train that departed at 10am carrying business and leisure travellers from both capital cities.

Flying Scotsman was displayed with the LNER dynamometer car along with GNR Stirling Single No. 1, GNR Atlantic No. 990 *Henry Oakley* and Deltic D9002 *King's Own Yorkshire Light Infantry* lined up in chronological order as supporting acts.

THE FIRST THREE YEARS

Fortunately, in view of the serious lineside trespassing problem encountered on its run from King's Cross, *Flying Scotsman* was not immediately scheduled for any further main line action and it was certainly questionable whether it would be made welcome on the East Coast Main Line again.

In March, the engine made its way to The North Yorkshire Moors Railway for a seven days' operation between the 12th and the 20th. Every train was sold out and the crowds flocked to the lineside to catch a glimpse of the engine. With eight coach trains and gradients as steep as 1-in-49, no chances were taken and another engine was attached to the rear of the train to give assistance from Grosmont as far as Goathland. After a couple of days though, the A3 was let loose on its own and gave the rare spectacle of a train passing through Goathland nonstop.

For main line operations it had been agreed that the engine would be used by railtour promoters Steam Dreams and the Railway Touring Company on a roughly equal basis and the two organisations planned very different programmes for the engine.

Steam Dreams got the first bite at the cherry on May 10 with a 'Cathedrals Express' from King's Cross, hauled by the Pacific from York to Newcastle and return.

The next train was somewhat more ambitious and although trespassing was less of a problem than it had been, this train hit the headlines for all the wrong, though completely different, reasons.

On May 14 another 'Cathedrals Express' set off from London and *Flying Scotsman* was again to take over the train at York but this time was booked to run right through to Edinburgh.

The previous September the Borders Railway had been opened between Edinburgh and Tweedbank on part of the old Waverley route to Carlisle via Galashiels. On the second day of the tour, it was planned to run *Flying Scotsman* over this new railway. However, the day before the train set off from King's Cross, the organisers were informed that *Flying Scotsman* was not 'gauged' to run to Tweedbank and would not therefore be able to do so.

Before any locomotive can run on any stretch of railway, Network Rail needs to check its computer records to confirm that it will not foul any bridges or other lineside structures. On this occasion this job had not been done despite the fact that sufficient notice had been given. Intensive lobbying of senior figures in the railway industry and in Scottish politics was immediately put into effect.

Fortunately, Network Rail was able to give the all-clear for *Scotsman* to haul the train to Tweedbank and over the Forth Bridge later in the day. This welcome news was received by the tour organisers while the train was en route to York.

Network Rail chief executive Mark Carne said: "Overnight and through today our engineers and analysts have worked hard to find a way to get the necessary safety checks and engineering assessments done.

"I am pleased to say that we have been successful and are now able to reinstate the original planned tours of *Flying Scotsman* in Scotland on Sunday.

"I wholeheartedly and sincerely apologise for the consternation caused by the premature announcement yesterday.

"Once the tours have been safely and successfully run, I will be instigating a full investigation into how this problem occurred on our railway in Scotland."

Scottish Transport Minister Derek Mackay had earlier accused Network Rail of "appalling incompetence" and described the situation as "a debacle".

"ScotRail have worked tirelessly over the past 24 hours and in that period of time have managed to sort out problems that Network Rail couldn't do in 12 weeks."

After the drama, *Flying Scotsman* made an uneventful run on May 17 from Edinburgh back to York.

No. 60103 then headed south for an intensive programme of full- and half-day dining trains still for Steam Dreams, on the Southern and Western Regions commencing on May 21 with a train from Paddington to Salisbury. Some trespassing was reported but operationally the engine was now hampered by some severe speed restrictions through various platforms in the London area.

The Railway Touring Company got a look-in on June 4 with *Scotsman* heading a railtour from Victoria to York one way, via Harringworth viaduct and the Midland Main Line through Loughborough. The engine then headed for Crewe to work the Crewe – Paddington leg of another Steam Dreams tour on June 8.

On June 11, the A3 headed an RTC tour from Cleethorpes to Newcastle and Bedlington and then on June 15 a Crewe – Holyhead trip for Steam

Dreams. June 18 saw the engine working part of a London – York tour for the RTC although it was not permitted to run on the southern section of the East Coast Main Line out of King's Cross. The engine was certainly touring the country but despite train timings being kept secret from the public, lineside trespass was a serious continuing problem.

On June 25, the Pacific ran one way from Victoria to York with an RTC train which was severely delayed by lineside trespassing en route. This positioned the engine at York for a summer programme of tours for RTC. These had originally been advertised as running over the Settle & Carlisle line but work was still progressing on repairing the landslip which had occurred in February when *Flying Scotsman* made its inaugural main line run to Carlisle and the line remained closed north of Appleby.

Instead 'The Hadrian' railtour on July 2 saw No. 60103 running from Carnforth over Shap to Carlisle and on to York via the Tyne Valley and Durham. This was followed by a series of seven 'Waverley' Sunday trips from York to Carlisle via Durham in both directions. All of the engine's main line runs continued to be fully booked often weeks in advance.

The autumn saw *Flying Scotsman* running more sedately on heritage lines, with a six-day visit to the Severn Valley Railway where it ran alongside the newly constructed LNER A1 Pacific No. 60163 *Tornado*, followed by a weekend's running on the East Lancashire Railway in October.

The engine spent the winter undergoing maintenance in Ian Riley's works on the East Lancashire Railway but started the 2017 operating season with a very important assignment. The landslip on the Settle & Carlisle line had been repaired and the line fully reopened to traffic in March. To mark the occasion, *Flying Scotsman* headed a railtour over the line on March 31 but this one was different in that it started at the Oxenhope terminus of the Keighley & Worth Valley Railway.

The A3 returned to the Worth Valley branch after the tour and spent the next nine days on the railway, heading passenger trains between Keighley and Oxenhope on most days.

April 13-19 saw *Scotsman* running in very unfamiliar territory, on the Bluebell Railway in Sussex, before returning to York for a very unusual trip. Very early on the morning of April 23, *Flying Scotsman* and a five-coach train of West Coast Railways' maroon Mk.1 coaches were towed backwards from York to Tollerton Junction a few miles to the north by the National Railway Museum's Deltic diesel D9002 *King's Own Yorkshire Light Infantry*.

A film was being made to promote the introduction of new 'Azuma' units on East Coast Main Line services. A quadruple parallel run was staged on the four-track ECML with *Flying Scotsman*, a new Azuma unit, a High-Speed Train and a Class 91 electric representing four eras of ECML train travel. The line had been closed overnight for engineering work so it was possible to stage this event, which was promoted by Virgin Trains, Welcome to Yorkshire, Network Rail and the National Railway Museum.

The four trains set off at sunrise on a perfect morning and ran parallel to Beningbrough where they stopped for a press photocall. Filmed from the air by helicopter, the trains set off and the Azuma accelerated away from the other three trains in the formation. The unique event was witnessed by a huge crowd of spectators who had got out of bed early.

The Railway Touring Company launched *Scotsman's* summer main line railtour programme on April 29 using the engine on the King's Cross – York leg of its annual 'Great Britain' tour. This was the RTC's tenth such tour which took in Kyle of Lochalsh, Fort William and Penzance using many different steam engines over a ten-day period.

It also marked *Flying Scotsman's* first departure from King's Cross since its official inaugural run of February 25 the previous year. Fortunately, the lineside trespassing issue was now easing off and no major problems were reported.

The main line programme took a very similar form to the previous year with Steam Dreams again promoting a tour to Scotland in May, although not taking in the Borders Railway this time. Steam Dreams promoted a series of dining trains in the south and west including runs to Minehead on the West Somerset Railway and from Victoria to Chester.

RTC did a Scarborough – King's Cross one way and a King's Cross – Loughborough – York run to position the engine for another series of summer Sunday 'Waverley' trips from York to Carlisle, this time running on the intended route over the Settle & Carlisle line.

Although *Scotsman* visited Minehead on a Steam Dreams' dining train, it had not hauled any West Somerset Railway trains. It returned to the railway in September for a four-day visit, returning to York on a 'Cathedrals Express' from Victoria on the 15th.

After a visit to Barrow Hill Roundhouse, on October 18, *Scotsman* set off from York to start a series of half-day dining trains for Steam Dreams in the east of England, an area which had not yet seen the engine.

En route to Peterborough via Lincoln it became evident that something was seriously wrong with the engine. It struggled into Peterborough at reduced speed and on arrival was pronounced unfit to carry on. While a diesel took over for the afternoon run from Ely, the A3 ran slowly round to Wansford on the Nene Valley Railway where repairs could be carried out.

A driving wheel bearing had overheated and the wheelset needed to be removed. This involved hiring two huge cranes to lift the locomotive clear of its driving wheels. In a major operation, the wheelset was removed and the bearing removed and repaired by Riley Engineering at Bury.

After running-in on the NVR, the engine was ready to return to service on the RTC's November 4 Ealing Broadway-to-York railtour. Although Steam Dreams' East Anglian programme was much truncated, *Scotsman* was still able to head a Norwich-to-Ipswich train followed by a Norwich-to-King's Cross train on November 11 before retiring for winter maintenance.

For 2018, the National Railway Museum again announced details of *Flying Scotsman*'s planned travels around Britain in the spring. The NRM's contract with Riley & Son (E) Ltd had initially been for two years and had now expired. As expected, the museum now signed a new six-year contract with the engineering firm to operate and maintain the engine.

Jim Lowe, Head of Operations at the National Railway Museum, said: "I would like to congratulate Riley & Son (E) Ltd for their success in winning the contract and I look forward to working with the team again in what is sure to be a very popular 2018 touring schedule. *Flying Scotsman* is a true symbol of engineering excellence and continues to inspire and amaze crowds of people wherever it goes.

"*Flying Scotsman* is a fantastic ambassador for the National Railway Museum and we aim to give as many people as possible the chance to see this legend of the steam age."

As well as day-to-day operation and maintenance, the contract will include a comprehensive overhaul in 2022 and plans to mark the engine's centenary celebrations in 2023.

Ian Riley, from Riley & Son (E) Ltd, said: "I am very pleased that we have been appointed to maintain and operate *Flying Scotsman* for the next six years which will enable us to develop a long-term plan for the care and management of this world-renowned engine. The success is down to the skills, experience and hard work of the whole team who have spent many hours working on the locomotive from restoration to the present day. *Flying Scotsman* is definitely a special engine and the chance to work with such a well-loved locomotive again is a real privilege."

Marcus Robertson, founder of Steam Dreams, said: "To start our *Flying Scotsman* season with the iconic trip to Edinburgh is always magical but then finishing with the loco on the first leg of a holiday in The Lakes will be the icing on the cake. It is hard to describe the emotions this amazing loco seems to evoke, but we anticipate a very popular public response once again."

Nigel Dobbing, Managing Director of The Railway Touring Company, said: "We are delighted to have again secured the services of *Flying Scotsman* to haul some of our tours this year, including the first leg of our 'Great Britain XI' tour. This iconic steam locomotive is always a very popular addition to our programme of daytrips within the UK."

East Lancashire Railway Chairman, Mike Kelly, said: "We are very proud of our relationship with *Flying Scotsman*. From her debut visit to the ELR back in 1993, to being the first heritage railway to host her following her extensive £4.2million restoration in January 2016, to now where she will make a number of appearances on our line throughout the year, captivating thousands more visitors as the world's most famous engine. She has a very special place in our hearts and we can't wait to welcome her back to the North West."

Now that a longer-term contract had been agreed, the engine was likely to spend considerable periods on the East Lancashire Railway. Accordingly, the first opportunity for the public to see *Flying Scotsman* was at the ELR during March, where it was on static display at Rawtenstall during the railway's steam gala and working ELR services on March 12/13.

The engine's main line programme was to be similar to the previous two years and commenced on April 19 with the King's Cross – Scarborough leg of the Railway Touring Company's 'Great Britain XI'. Next came Steam Dreams' trip to Scotland over May 19-22 working York to Edinburgh and three Fife Circle trips but again no run on the Borders Railway and returning to York.

This was followed by further Steam Dreams dining trains in the southeast and a couple of RTC trips plus three Settle & Carlisle trains during the summer before returning to the ELR for August 24-27.

Flying Scotsman visited the Nene Valley Railway as a special thank-you for assistance given after its failure the previous year, and ran on the line over September 29 – October 1. At the time, the railway was playing host to the LNER A1 Pacific No. 60163 *Tornado* which had also suffered a major failure on an ECML railtour resulting in its being taken to Wansford for major repairs. These repairs had taken very much longer than *Scotsman*'s did, but the A1 was now accumulating considerable running-in mileage.

Flying Scotsman also made an appearance at the National Railway Museum in York and at the Locomotion museum in Shildon.

In early October, No. 60103 made its first visit to Cornwall, heading a series of Steam Dreams dining trains in the West Country, one of which took the engine to Penzance. Most of these trains were double-headed with one of Ian Riley's LMS 'Black Five' 4-6-0s, No. 44871.

On October 13, Steam Dreams ran *Flying Scotsman* from King's Cross to York as a tribute to Alan Pegler who had purchased the engine from British Railways in 1963. Alan had died on March 18, 2012 at the age of ninety-one. His ashes were placed in the firebox of the locomotive on the climb of Stoke Bank.

In a new development, Ian Riley promoted a couple of railtours in the late autumn and *Flying Scotsman* was booked for one of these, running over the Settle & Carlisle line and returning over Shap on December 22.

The National Railway Museum and the rail industry came together on Friday 11 January to honour the memory of Sir William McAlpine who had rescued *Flying Scotsman* from America in 1973 and operated it for the next 23 years. Sir William had passed away earlier on March 4, 2018 at the age of 82 and to recognise his significant contribution to railway heritage a special memorial trip hauled by *Flying Scotsman* was run in his memory.

As a further tribute, D.B. Cargo, Network Rail and LNER renamed a Class 90 electric locomotive in his memory. Named 'the Scotsman's Salute', the train ran from King's Cross to York on Friday, 11 January, 2019, which would have been Sir William's birthday. The naming ceremony was held at the National Railway Museum and the newly named locomotive *Sir William McAlpine* then hauled the memorial trip's return journey to King's Cross.

As a final tribute, money from the sale of each ticket will be used to fund a one-year engineering traineeship. One lucky budding engineer will get the chance to travel around the country with custodians Riley & Son (E) learning how to maintain the famous engine.

Jim Lowe, Head of Operations at the National Railway Museum, said: "On behalf of myself, colleagues and volunteers, I would like to publicly thank Sir William for his significant contribution to the museum and to the wider preservation of railway heritage in this country. Holding this memorial tour, naming a locomotive after him and especially, setting up an engineering bursary to benefit young people, will create a fitting legacy to honour his memory."

Time has flown since the legend that is *Flying Scotsman* was reborn and the engine returned to steam at the end of 2015. The Gresley Pacific is already half way through its seven-year period of certification for main line operation. It continues to travel the country reliably with only the occasional hiccup.

Thoughts will now be turning towards keeping the engine going for as long as possible. After seven years has elapsed, it is expected to be immediately withdrawn from service for a major overhaul to return it to the main line for a further period, including the celebration of its 100th birthday. The engine's popularity shows no sign of diminishing and it continues to attract huge crowds wherever it goes.

Appendix 1:

THE LOCOMOTIVE ENGINEERS

Great Northern Railway Locomotive Superintendents
  - » Archibald Sturrock 1850-1866
  - » Patrick Stirling 1866-1895
  - » Henry Alfred lvatt 1896-1911
  - » Herbert Nigel Gresley 1911-1922

London & North Eastern Railway Chief Mechanical Engineers
  - » Herbert Nigel Gresley 1923-1941
  - » Edw.ard Thompson 1941-1946
  - » A.H Peppercorn 1946-1948

Some of Gresley's Contemporary Locomotive Engineers in 1911
  - » Great Western Railway: G.J. Churchward 1902-1921
  - » London & South Western Railway: Dugald Drummond 1895-1912
  - » London & North Western Railway: C.J. Bowen – Cooke 1909-1920
  - » Midland Railway: Henry Fowler 1909-1922
  - » Great Central Railway: J.G. Robinson 1900-1922
  - » Great Eastern Railway: S.D. Holden 1908-1912
  - » North Eastern Railway: Sir Vincent Raven 1910-1922
  - » North British Railway: W.P. Reid 1903-1919

Appendix 2:

GNR/LNER LOCOMOTIVE CLASSES

» A. 4-6-2 'Pacific': including original A1 such as *Flying Scotsman*, and the A3 developments, NER Raven A2 and later Gresley's streamlined A4. Also, GCR and NER 4-6-2 tank engines.

» B. 4-6-0: including Gresley's B17 'Sandringhams', but Thompson's B1 was the most common.

» C. 4-4-2: 'Atlantic', built by most pre-Grouping constituents of the LNER, but mostly withdrawn by early BR days, apart from tank engines.

» D. 4-4-0: including the GER Claud Hamiltons, GCR Directors, and NBR 'Glens', and Gresley's LNER D49 'Hunts & Shires'.

» E. 2-4-0: The GER E4s ran well into BR days in the 1950s.

» F. 2-4-2: mostly tank engines.

» G. 0-4-4: mostly tank engines.

» H. 4-4-4: The NER built a few, as did the Metropolitan Railway.

» J. 0-6-0: of numerous varieties, including early diesel shunters.

» K. 2-6-0: Gresley's GNR K2 and K3, and LNER K4, plus Thompson's later K1.

» L. 2-6-4: all were tank engines.

» M. 0-6-4: a few tank engines.

» N. 0-6-2: mostly tank engines, including Ivatt's N1 and Gresley's N2 for King's Cross suburban trains.

» 0. 2-8-0: including the OCR 04, adopted by the WD in the First World War, and Gresley's first three-cylinder design for the GNR.

» P. 2-8-2: The two P1s were Gresley's big LNER goods engines, but the P2s were his biggest express engines, such as No. 2001 *Cock o' the North*.

» Q. 0-8-0: large goods engines, especially on the NER and GNR.

» R. 0-8-2:

» S. 0-8-4:

» T. 4-8-0: U: The unique Garratt, No. 9999, a 2-8-8-2.

» U. Beyer-Garratt articulated locomotives; the LNER had just one, built during Gresley's reign.

» V. 2-6-2: rare in Britain, but Gresley's V2 'Green Arrows' were exceptional.

» W. 4-6-4: again rare, but Gresley built one; the experimental water-tube boilered No. 10000. Technically it was, in fact, a 4-6-2-2.

» X. 2-2-4 and 4-2-2.

» Y. 0-4-0 small tank engines.

» Z. 0-4-2.

» The LNER locomotive classifications continued in use throughout BR days, for ex-LNER engines.

Appendix 3:

THE A1s AND A3s

Legend: The engines are shown in chronological order of construction, from 1922 to 1935. The LNER numbering system was confusing and, when engines were renumbered, they were not kept in their original order at all, so the eventual BR numbering of the class is really quite haphazard, with the original engines carrying the highest numbers. Not all of the engines were built at Doncaster, Nos 2563 to 2582 being built by the North British Locomotive Company in Glasgow.

A1				BR no.
1470	4470	*Great Northern*	Rebuilt as Thompson prototype A1 1945	60113
1471	4471	*Sir F. Banbury*	A3 from 1942	60102
1472	4472	*Flying Scotsman*	A10 from 1945: A3 from 1947	60103
1473	4473	*Solario*	A3 from 1941	60104
1474	4474	*Victor Wild*	A3 from 1942	60105
1475	4475	*Flying Fox*	A10 from 1945: A3 from 1947	60106
1476	4476	*Royal Lancer*	A10 from 1945: A3 from 1946	60107
1477	4477	*Gay Crusader*	A3 from 1943	60108
1478	4478	*Hermit*	A3 from 1943	60109
1479	4479	*Robert the Devil*	A3 from 1942	60110
1480N	4480	*Enterprise*	A3 from 1927	60111
1481N	4481	*St Simon*	A10 from 1945: A3 from 1946	60112
	2543	*Melton*	A10 from 1945: A3 from 1947	60044
	2544	*Lemberg*	A3 from 1927	60045
	2545	*Diamond Jubilee*	A3 from 1941	60046
	2546	*Donovan*	A10 from 1945: A3 from 1947	60047
	2547	*Doncaster*	A10 from 1945: A3 from 1946	60048

2548	Galtee More	A10 then A3 from 1945	60049
2549	Persimmon	A3 from 1943	60050
2555	Blink Bonny	A10 then A3 from 1945	60051
2551	Prince Palatine	A3 from 1941	60052
2552	Sansovino	A3 from 1943	60053
2553	Prince of Wales	A3 from 1943 (previously *Manna*)	60054
2554	Woolwinder	A3 from 1942	60055
2555	Centenary	A3 from 1944	60056
2556	Ormonde	*A10 from 1945: A3 from 1947*	60057
2557	Blair Athol	A10 then A3 from 1945	60058
2558	Tracery	A3 from 1942	60059
2559	The Tetrarch	A3 from 1942	60060
2560	Pretty Polly	A3 from 1944	60061
2561	Minoru	A3 from 1944	60062
2562	Isinglass	A10 from 1945: A3 from 1946	60063
2563	Tagalie	A3 from 1942 (*William Whitelaw*)	60064
2564	Knight of Thistle	A10 from 1945: A3 from 1947	60065
2565	Merry Hampton	A10 then A3 from 1945	60066
2566	Ladas	A3 from 1939	60067
2567	Sir Visto	A10 from 1945: A3 from 1948	60068
2568	Sceptre	A3 from 1942	60069
2569	Gladiateur	A10 from 1945: A3 from 1947	60070
2570	Tranquil	A3 from 1944	60071
2571	Sunstar	A3 from 1941	60072
2572	St Gatien	A10 then A3 from 1945	60073
2573	Harvester	A3 from 1928	60074
2574	St Frusquin	A3 from 1942	60075
2575	Galopin	A3 from 1941	60076
2576	The White Knight	A3 from 1943	60077
2577	Night Hawk	A3 from 1944	60078
2578	Bayardo	A3 from 1928	60079
2579	Dick Turpin	A3 from 1942	60080
2580	Shotover	A3 from 1928	60081
2581	Neil Gow	A3 from 1943	60082
2582	Sir Hugo	A3 from 1941	60083

Engines built as A3 from 1928

2743	*Felstead*	60089
2744	*Grand Parade*	60090
2745	*Captain Cuttle*	60091
2746	*Fairway*	60092
2747	*Coronach*	60093
2748	*Colorado*	60094
2749	*Flamingo*	60095
2750	*Papyrus*	60096
2751	*Humorist*	60097
2752	*Spion Kop*	60098
2595	*Trigo*	60084
2596	*Manna*	60085
2597	*Gainsborough*	60086
2795	*Call Boy*	60099
2796	*Spearmint*	60100
2797	*Cicero*	60101
2598	*Blenheim*	60087
2599	*Book Law*	60088
2500	*Windsor Lad*	60035
2501	*Colombo*	60036
2502	*Hyperion*	60037
2503	*Firdaussi*	60038
2504	*Sandwich*	60039
2505	*Cameronian*	60040
2506	*Salmon Trout*	60041
2507	*Singapore*	60042
2508	*Brown Jack*	60043

Index

North Eastern Region (BR's) 49, 68
North Staffordshire Railway 25, 39
North Yorkshire Moors Railway 142
North British Waverley 55
North London Railway 25
Norwich 51, 113, 115, 146
Nottingham 15

Orient Express *see* Venice Simplon Orient-Express
Oxenhope terminus 144
Oxford Molecular Group plc 106, 117, 118

Pacifics 19, 20, 22, 23, 24, 28-32, 35, 37, 48, 42-47, 50, 53, 55, 56, 59, 62, 63, 66, 69, 70, 77, 84, 90, 92, 93, 107-110, 119, 128, 132, 142, 144, 148, 149
Paddington station 34, 40, 59, 143
Parkes to Broken Hill record 93, 103
Pegler, Alan 11, 57, 58, 59, 61, 63, 65-69, 71, 73, 74, 84, 88, 96, 110, 112, 124-126, 131, 132, 140, 148
Paignton & Dartmouth Steam Railway *see* Torbay Steam Railway
Penmanshiel Tunnel collapse 87
Pennines 59, 129
Penrith station 137
Penydarren tramway locomotive 120
Penzance 34, 145, 148
Peppercorn, A.H. *also* Pacifics 26, 47, 50, 54, 109
Perth, Scotland 81
Perth, WA 92, 94
Peterborough 14, 15, 27, 42, 44, 55, 58-60, 88, 89, 131, 139, 146
Plymouth 34, 81, 115
Pole, Sir Felix 31
Portillo, Michael 139
privatisation (of BR) 82, 94-100, 108
Proctor, Dave 139

'Races to the North' 14
Rail Express Systems 98, 99
Railfest 120, 130
Railfreight 98

railtours 56, 65, 76, 83, 88, 98, 99, 136, 137, 142-146, 148
Railtrack 79, 98, 108
Railway Touring Company 88, 136, 142, 143, 145, 147, 148
Raven, Sir Vincent 19, 30, 45
Ravenglass & Eskdale miniature railway 83
Rees, Jim 121, 122, 124
Regional Railways 98
Retford 14, 27, 66, 109, 139
Ribble Valley 136
Ribblehead 137
Richards, Les 68
Riddles, Robert 50
Riley & Son Engineering 125, 129, 130, 134, 135, 146, 147, 149
Riley, Ian 135, 147, 148
Robertson, Marcus 147
Rocket 22, 24, 62
Royal Trains 89
Rugby 14

Salisbury 81, 90, 143
San Francisco 70, 71
San Francisco Belt Railroad 70
Save Our Scotsman campaign 118, 119, 129
Scarborough 35, 77, 81, 82, 85, 86, 110, 121, 122, 126, 127, 137, 138, 145, 148
Scott, Andrew 129
Scottish Region (BR's) 49
Sea Containers Ltd 110
Sellafield *also* British Nuclear Fuels 81, 83, 84
Severn Valley Railway 64, 94, 97, 144
Shap 61, 83, 87, 136, 137, 144, 148
Sharpe, Brian 65, 124
Shildon *see also* Locomotion museum 65, 81, 85, 121, 148
Shrewsbury 76, 81, 115
Sir Robert McAlpine Ltd 73
Smith, David 82
Smith, Jim 139
South Eastern & Chatham Railway 25, 38